SpringerBriefs in Mathematics

Series Editors

Krishnaswami Alladi
Nicola Bellomo
Michele Benzi
Tatsien Li
Matthias Neufang
Otmar Scherzer
Dierk Schleicher
Benjamin Steinberg
Vladas Sidoravicius
Yuri Tschinkel
Loring W. Tu
G. George Yin
Ping Zhang

SpringerBriefs in Mathematics showcases expositions in all areas of mathematics and applied mathematics. Manuscripts presenting new results or a single new result in a classical field, new field, or an emerging topic, applications, or bridges between new results and already published works, are encouraged. The series is intended for mathematicians and applied mathematicians.

For further volumes:
http://www.springer.com/series/10030

Joanna A. Ellis-Monaghan • Iain Moffatt

Graphs on Surfaces

Dualities, Polynomials, and Knots

 Springer

Joanna A. Ellis-Monaghan
Department of Mathematics
Saint Michael's College
Colchester, VT 05439
USA

Iain Moffatt
Department of Mathematics
Royal Holloway University of London
Egham, Surrey TW20 0EX
UK

ISSN 2191-8198   ISSN 2191-8201 (electronic)
ISBN 978-1-4614-6970-4   ISBN 978-1-4614-6971-1 (eBook)
DOI 10.1007/978-1-4614-6971-1
Springer New York Heidelberg Dordrecht London

Library of Congress Control Number: 2013933997

Mathematics Subject Classification (2010): 05C10, 05C31, 57M15, 57M27

Printed on acid-free paper

Springer is part of Springer Science+Business Media (www.springer.com)

Preface

This monograph aims to provide an accessible and fairly comprehensive treatment of recent developments on generalised dualities for graphs on surfaces and their applications. Duality arises in many areas, particularly topological and algebraic graph theory, topology, and physics. The importance of duality derives not only from its inherent properties but also from its interactions with functions on graphs (such as graph polynomials) and knot invariants. Traditionally, most of graph theory concerning polynomials and knot invariants has focused on properties of abstract or plane graphs. However, new research has impelled an analogous theory for graphs on surfaces. Here we examine the full generalisation of duality for embedded graphs and the interactions of this duality with graph polynomials and knot polynomials that resulted from this research. We illustrate some of the advantages of moving from plane and abstract graphs to graphs on surfaces. Although primarily a survey, this text does give new approaches to the material and contains several new results.

We focus on four key interdependent topics:

- Extending geometric duality fully to graphs on surfaces
- Properties of 4-regular graphs and medial graphs
- Relations, identities, and interpretations for polynomials of graphs on surfaces
- Connections between knot theory and graph theory

We begin by briefly cataloguing various descriptions of graphs on surfaces and reviewing the classical constructions of the Petrie dual, $G^\times$, geometric dual, G^*, and the medial graph, G_m, of an embedded graph G. This leads to our primary motivation, namely the classical relationships among the medial graphs and the geometric duals of plane graphs. Suppose that G is a plane graph with dual G^* and medial graph G_m. The medial graph of G^* is exactly the medial graph of G, i.e., $(G^*)_m = G_m$, where $=$ denotes equality as plane graphs. In fact, the connection between geometric duals and medial graphs is a little stronger than this. The two graphs G and G^* are the only plane graphs that have G_m as their plane medial graphs, that is,

$$\{G, G^*\} = \{H \,|\, H_m = G_m\}. \tag{1}$$

Twisted duality extends the fundamental classical relations among a plane graph, its plane dual, and its medial graph to graphs embedded in arbitrary surfaces. It arises from "localising" the classical constructions of geometric and Petrie duals to individual edges. These local operations lead to a group action, called the ribbon group action, on the set of embedded graphs. The twisted duals of an embedded graph comprise the orbit of it under this group action. Other forms of duality, such as geometric duality, Petriality, and partial duality, appear as actions of different subgroups of the ribbon group. Twisted duality gives a full surface analogue of Eq. (1) in that if G is any embedded graph with medial graph G_m, then its twisted duals are precisely the set of all embedded graphs with medial graphs isomorphic (as abstract graphs) to G_m. Furthermore, in analogy with how a plane graph and its dual may be reconstructed from the medial graph, all the twisted duals of an embedded graph may be constructed from its medial graph. Exploring Eq. (1) further, we find that the type of graph duality on the left-hand side and the type of graph isomorphism on the right-hand side are inextricably connected in that substituting another kind of graph isomorphisms for $=$ in the left-hand side of Eq. (1) corresponds to a particular form of duality on the right-hand side. Isomorphism as embedded graphs corresponds to geometric duality, and isomorphism as abstracts graphs corresponds to twisted duality, but further, we show that twisted duality gives a hierarchy of various forms of graph duality from the literature that correspond, through appropriate analogues of Eq. (1), to a hierarchy of graph isomorphism.

After establishing twisted duality, we turn to its interactions with invariants of graphs on surfaces, particularly graph polynomials. Recently, several graph polynomials that were originally defined for abstract or plane graphs have been extended to graphs on surfaces. These include the transition polynomial, the Penrose polynomial, and several different extensions of the Tutte polynomial. Our main tool here is the topological transition polynomial, which interacts naturally with the ribbon group action and coincides with these other polynomials as well as the Kauffman bracket of knot theory. The ribbon group action leads to new properties of the transition polynomial and from there to a deeper understanding of the properties of, and relationships among, various graph polynomials. The advantages of this approach are particularly well illustrated by the Penrose polynomial. The Penrose polynomial of a plane graph, which encodes colouring information, first appeared implicitly in the work of R. Penrose on diagrammatic tensors. Extending the Penrose polynomial to graphs on surfaces, and using its relation to the transition polynomial, reveals many new properties that simply cannot be realised in the original plane setting. These include deletion–contraction reductions, duality relations, and a restatement of the Four Colour Theorem. Similarly, we show how the ribbon group action and its interaction with the transition polynomial lead to new identities for the ribbon graph and topochromatic polynomials. These identities for graph polynomials subsequently inform applications to knot theory.

Generalising duality to graphs on surfaces has its origins in knot theory. S. Chmutov and I. Pak showed that the Jones polynomial of an alternating checkerboard colourable virtual link diagram is an evaluation of the ribbon graph polynomial of B. Bollobás and O. Riordan (which generalises the Tutte polynomial from abstract to embedded graphs). This result extends a seminal theorem of M. Thistlethwaite which relates the Tutte polynomial of a plane graph and the Jones polynomial of an alternating (classical) link. Chmutov and Pak's paper stimulated considerable research into connections among polynomials of embedded graphs, knot polynomials, and representations of link diagrams as embedded graphs. To connect the various realisations of the Jones polynomial as a graph polynomial, Chmutov introduced an extension of geometric duality called partial duality, one of the inspirations of twisted duality. We use the theory described in this text to unify various connections among dualities, graph polynomials, and knot polynomials. We emphasise the ways in which developments in knot theory lead to developments in graph theory, and vice versa, and take the reader to the forefront of research in this area.

Fundamentally, this text illustrates the interdependency between duality, medial graphs, and knots; how this interdependency is reflected in algebraic invariants of graphs and knots; and how this interdependency can be exploited to solve problems in graph theory and knot theory. Throughout, we take a constructive approach, emphasising how the ideas and constructions described here arise from classical constructions such as geometric duals and Tait graphs, by removing artificial restrictions in these constructions, by localising global operations, or by broadening the setting to embedded graphs. We describe how these adaptations may be accomplished and what the benefits of doing so are.

Our goal is to give a self-contained introduction to graphs on surfaces, twisted duality, and topological graph and knot polynomials that is accessible to both graph theorists and knot theorists. Accordingly, we have assumed familiarity with only basic graph theory and knot theory so that the text should be accessible to graduate students and researchers in either area. Because the area is advancing so rapidly, we have not attempted an exhaustive catalogue but rather tried to give a comprehensive overview, with a robust bibliography, hoping to provide the reader with the necessary knowledge and background to read research papers on these topics as they appear. We hope that the reader will come away from the text convinced of advantages of considering these higher genus analogues of constructions of plane and abstract graphs and with a good understanding of how they arise.

As a final remark, Chaps. 1 and 2 contain the common foundational material for Chaps. 3, 4 and 5. Chapters 4 and 5 do not depend upon Chap. 3 (so a reader interested only in graph polynomials or knots may safely skip Chap. 3). Chapter 5 uses material from Sects. 4.1 to 4.3 and 4.5.

Colchester, VT
Egham, UK
Joanna A. Ellis-Monaghan
Iain Moffatt

Contents

Chapter 1
Embedded Graphs

There are several different methods for describing graphs embedded in surfaces, each with its own advantages. We begin by reviewing cellularly embedded graphs and their representations as ribbon graphs (the form in which we will generally work with them), band decompositions, ram graphs, arrow presentations, and signed rotation systems. We also describe Petrie duals, geometric duals, medial graphs and Tait graphs, and the relations among them, as these definitions and relations motivate much of the work presented later in the monograph.

Our aim in this section is to provide a quick and concise introduction to various realisations of embedded graphs and how to move between them. Accordingly, we do not attempt to give a full rigorous treatment of the equivalences among the various representations of embedded graphs. Instead we work somewhat intuitively, providing geometric rationales rather than formal proofs for the equivalences. We refer the reader to, for example, [41, 81] for a rigorous treatment of embedded graphs.

1.1 Embedded Graphs and Their Representations

We assume the reader has some familiarity with basic graph theory, surface topology, and graphs embedded in surfaces. See, for example, [7, 24, 41, 64, 81, 98] for good expositions on graph theory and topological graph theory.

1.1.1 Abstract Graphs

Graphs may have multiple edges and loops. We use standard notation. If G is a graph, then $V(G)$ is its vertex set and $E(G)$ its edge set. Also, $v(G)$, $e(G)$, and $k(G)$, respectively, are the numbers of vertices, edges, and components of G. The

J.A. Ellis-Monaghan and I. Moffatt, *Graphs on Surfaces: Dualities, Polynomials, and Knots*, SpringerBriefs in Mathematics, DOI 10.1007/978-1-4614-6971-1_1,
© Joanna A. Ellis-Monaghan, Iain Moffatt 2013

rank of G is $r(G) := v(G) - k(G)$, and the *nullity* of G is $n(G) := e(G) - r(G)$. The nullity, $n(G)$, of a graph is also called the *cycle rank, circuit rank*, the *Betti number*, or the *1st-Betti number* of a graph and is sometimes denoted by $\beta(G)$ or $\beta_1(G)$. The number of components $k(G)$ is often denoted by $\beta_0(G)$ and is sometimes called the *0th-Betti number*.

A subgraph $H = (U, A)$ of $G = (V(G), E(G))$ is a *spanning subgraph* if $U = V(G)$. If U consists of exactly the vertices of G that are incident to edges in $A \subseteq E(G)$, H is the *subgraph induced by A*, denoted $G|_A$. The complement of $A \subseteq E(G)$ is $A^c := E(G) \backslash A$. Given a graph G and $A \subseteq E(G)$, we let $e(A)$, $v(A)$, $k(A)$, $r(A)$, and $n(A)$ denote the number of edges, vertices, components, rank, and nullity of the spanning subgraph $(V(G), A)$ of G.

A walk traverses consecutive edges in a graph, allowing repeated edges and vertices; a trail allows repeated vertices but not edges; a path allows neither. A circuit is a closed trail, and a cycle is a closed path. A graph is *Eulerian* if it contains a circuit that visits every edge exactly once. A connected graph is Eulerian if and only if the degree of each of its vertices is even. A graph is *bipartite* if each of its vertices can be assigned one of two colours in such a way that adjacent vertices are assigned different colours (i.e., it is 2-colourable). It is well known that a graph is bipartite if and only if it contains no cycles of odd length.

Because our main focus is on embedded graphs, we will generally use the term "abstract graph" to emphasise that the graph is not embedded. Similarly, we will refer to two isomorphic graphs G and H as being *isomorphic as abstract graphs*, writing $G \cong H$ for this isomorphism.

1.1.2 *Surfaces*

In this section we briefly review the classification of surfaces. We refer the reader to, for example, [41, 73] for full details of the classification of surfaces.

A mapping between topological spaces is a *homeomorphism* if it is continuous and has a continuous inverse. A *surface* Σ is a compact topological space in which distinct points have distinct neighbourhoods, and each point has a neighbourhood that is homeomorphic to either the plane $\mathbb{R}^2$ or the upper half-plane $\mathbb{R}^2_+$. If each point has a neighbourhood homeomorphic to the plane, Σ is a *closed surface*, otherwise it is called a *punctured surface* or a *surface with boundary*. Note that our surfaces need not be connected and that we have defined our surfaces to be compact. Surfaces are considered up to homeomorphism.

Some surfaces are shown in Fig. 1.1. The real projective plane, $\mathbb{R}P^2$, is shown in Fig. 1.1c. It is obtained by identifying the two sides of a disc D^2 in such a way that the directions of the arrows match up. Figure 1.1d shows Boy's surface, which is an immersion of $\mathbb{R}P^2$ in $\mathbb{R}^3$.

If Σ is a punctured surface, then a point p on Σ whose neighbourhoods are homeomorphic to the upper half-plane $\mathbb{R}^2_+$ is said to be a *boundary point*. The union of the boundary points of Σ is a collection of circles. These circles are the *boundary components* of Σ.

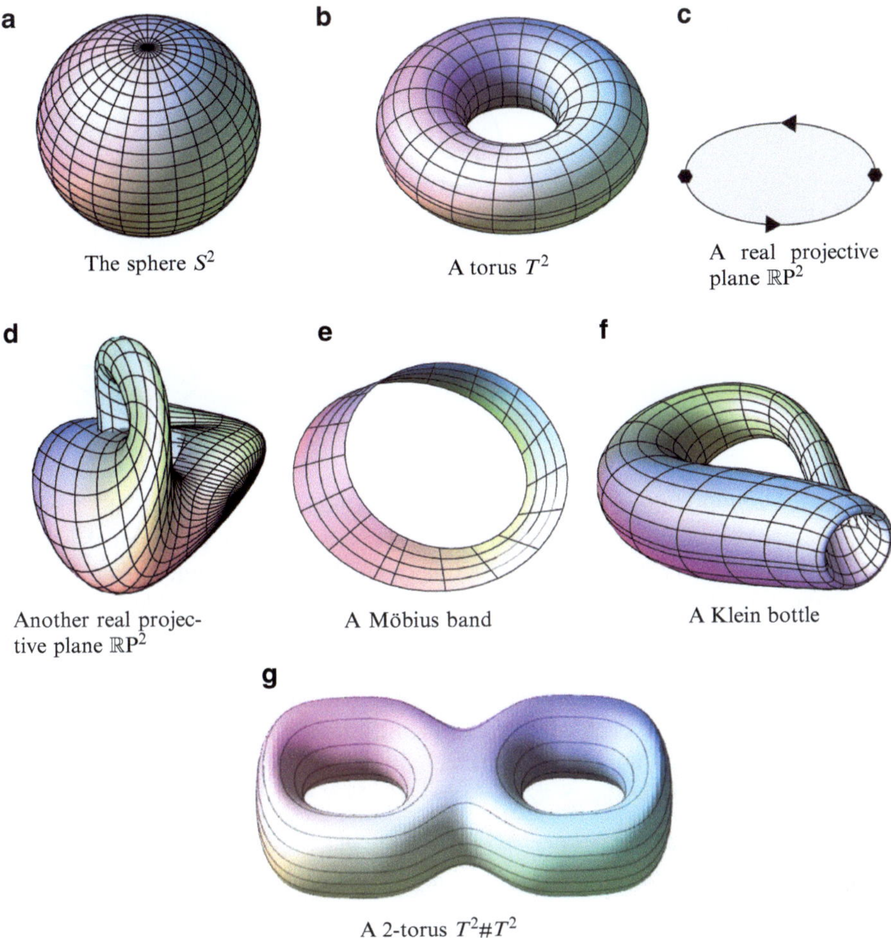

a The sphere S^2

b A torus T^2

c A real projective plane $\mathbb{RP}^2$

d Another real projective plane $\mathbb{RP}^2$

e A Möbius band

f A Klein bottle

g A 2-torus $T^2 \# T^2$

Fig. 1.1 Some examples of surfaces

For brevity, we will be somewhat intuitive in our discussion of orientability. At each point p of a surface Σ we can associate a concept of "left" or "right" by assigning a preferred direction of rotation (or a pair of orthogonal coordinate axes) to p, called a *local orientation* at p. If α is a path in Σ from p to another point q, then we can "move" the orientation of p along α to the point q. In particular, if p and q both have local orientations, then the orientation starts at p, moves around the surface in some way, then ends up at q. One of two things can happen: either the orientations agree, in which case the path α is said to be *orientation preserving*, or the orientations disagree, in which case α is said to be *orientation reversing*. (See Fig. 1.2 which shows an orientation reversing path on a Möbius band.) A surface is said to be *orientable* if every closed path is orientation preserving, and

Fig. 1.2 An orientation
reversing path on the Möbius
band

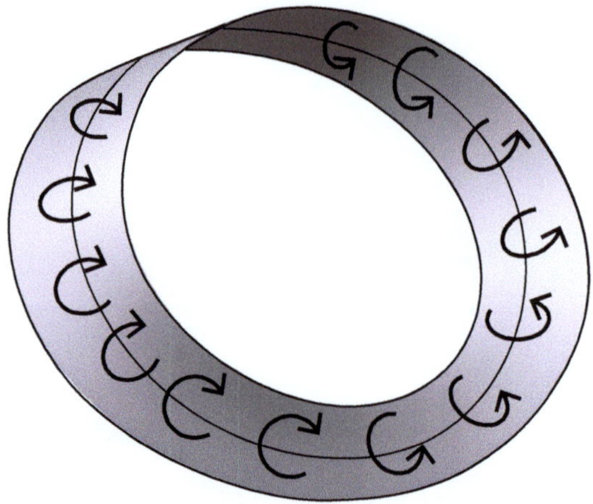

is *non-orientable* if there exists an orientation reversing closed path in the surface.
An orientable surface is said to be *oriented* if every point is assigned a local
orientation and every path in the surface is orientation preserving. In Fig. 1.1, the
real projective plane, Möbius band, and Klein bottle are non-orientable, while all of
the other surfaces shown are orientable.

The sphere, torus, and real projective plane form the building blocks of all
surfaces. The *connected sum*, $\Sigma\#\Sigma'$, of two surfaces Σ and Σ' is obtained by deleting
the interior of a disc in each surface and identifying the two boundaries that were
created. Figure 1.1g shows the connected sum of two tori. It can be shown that
$\mathbb{RP}^2\#\mathbb{RP}^2$ is the Klein bottle, $S^2\#\Sigma = \Sigma$ and $\mathbb{RP}^2\#T^2 = \mathbb{RP}^2\#\mathbb{RP}^2\#\mathbb{RP}^2$.

Theorem 1.1. *Let Σ be a connected closed compact surface.*

1. *If Σ is orientable, then it is homeomorphic to either a sphere or a connected sum
   of tori.*
2. *If Σ is non-orientable, then it is homeomorphic to a connected sum of real
   projective planes.*

We can use the above theorem to define the genus, which is an invariant of a
surface. The *genus*, $g(\Sigma)$, of a closed surface Σ is defined by

$$
g(\Sigma) := \begin{cases} 0 & \text{if} \quad \Sigma \text{ is homeomorphic to } S^2; \\ n & \text{if} \quad \Sigma \text{ is homeomorphic to the connected sum of } n \text{ tori}; \\ n & \text{if} \quad \Sigma \text{ is homeomorphic to the connected sum of } n \text{ real projective planes}. \end{cases}
$$

Genus together with orientability provides a complete classification of closed
surfaces:

Theorem 1.2. *Let Σ_1 and Σ_2 be closed connected compact surfaces. Then Σ_1 and
Σ_2 are homeomorphic if and only if they are both orientable or both non-orientable,
and they have the same genus.*

We now describe the classification of punctured surfaces. Let Σ be a punctured surface. Each boundary component of Σ is homeomorphic to a circle. This means we can obtain a closed surface Σ' from Σ by identifying each boundary component of Σ with the boundary of a (distinct) disc D^2. We say that the closed surface Σ' is obtained from Σ by *capping off the punctures*.

The *genus of a punctured surface* is defined to be the genus of the closed surface obtained by capping off each of the punctures.

Theorem 1.3. *Let Σ_1 and Σ_2 be connected compact punctured surfaces, and Σ_1' and Σ_2' be the closed surfaces obtained by capping off the punctures. Then Σ_1 and Σ_2 are homeomorphic if and only if they have the same number of boundary components, and Σ_1' and Σ_2' are homeomorphic.*

Genus, orientability, and the number of boundary components provide a complete classification of closed surfaces:

Theorem 1.4. *Two connected compact punctured surfaces are homeomorphic if and only if they have the same number of boundary components, the same genus and are either both orientable or both non-orientable.*

1.1.3 Cellularly Embedded Graphs

A *cellularly embedded graph* $G = (V(G), E(G)) \subset \Sigma$ is a graph drawn on a surface Σ in such a way that edges only intersect at their ends and such that each connected component of $\Sigma \backslash G$ is homeomorphic to a disc. If G is cellularly embedded in Σ, the connected components of $\Sigma \backslash G$, when viewed as subsets of Σ, are called the *faces* of G. Figure 1.3a shows a graph cellularly embedded in the torus.

Two cellularly embedded graphs $G \subset \Sigma$ and $G' \subset \Sigma'$ are *equivalent*, written $G = G'$, if there is a homeomorphism (which is orientation preserving when Σ is orientable) from Σ to Σ' that sends G to G'. That is, there is a homeomorphism $\varphi : \Sigma \to \Sigma'$ (which is orientation preserving when Σ is orientable) with the property that $\varphi|_G : G \to G'$ is an isomorphism of abstract graphs. We consider cellularly embedded graphs up to this equivalence.

1.1.4 Ribbon Graphs

Ribbon graphs are equivalent to cellularly embedded graphs, but they have the advantage over cellularly embedded graphs that deleting edges or vertices of a ribbon graph results in another ribbon graph, whereas deleting an edge of a cellularly embedded graph may not result in a cellularly embedded graph. We will often work in the language of ribbon graphs for this reason.

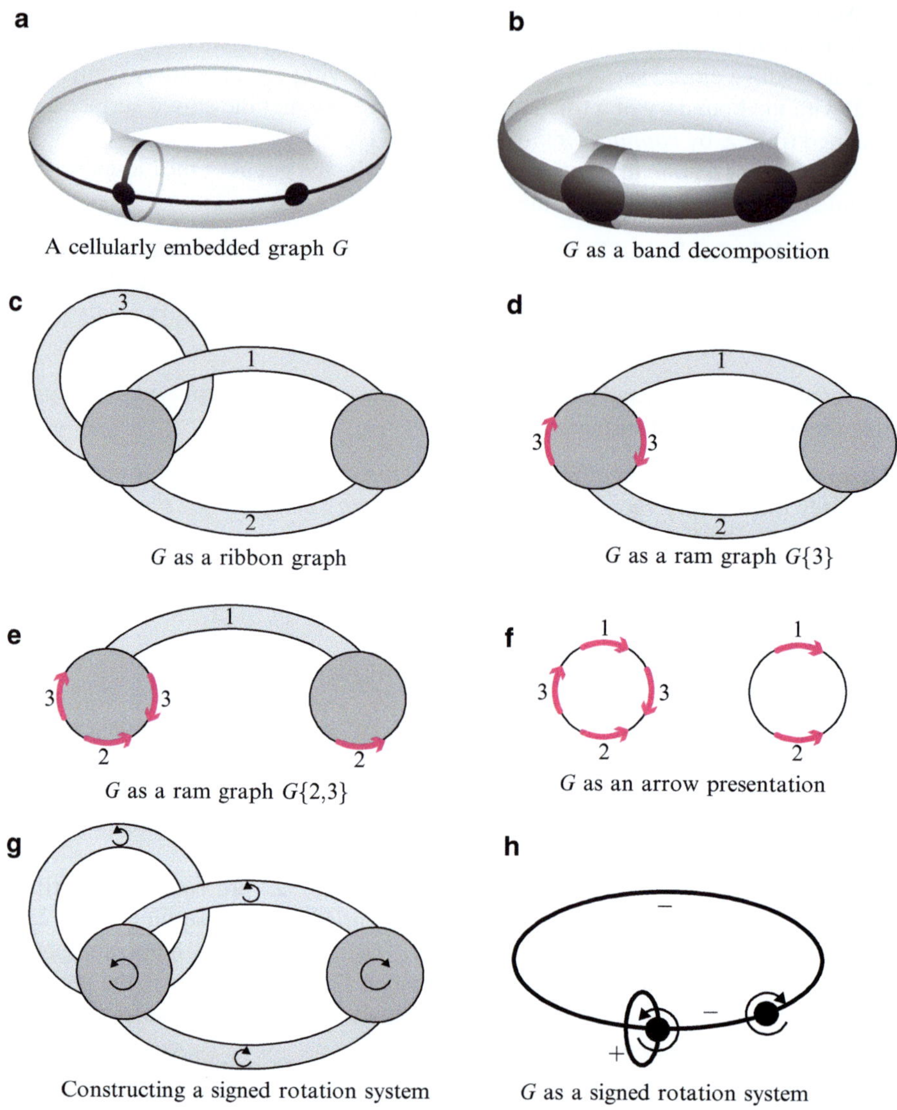

a A cellularly embedded graph G

b G as a band decomposition

c G as a ribbon graph

d G as a ram graph $G\{3\}$

e G as a ram graph $G\{2,3\}$

f G as an arrow presentation

g Constructing a signed rotation system

h G as a signed rotation system

Fig. 1.3 Different presentations of the same embedded graph

Definition 1.5. A *ribbon graph* $G = (V(G), E(G))$ is a (possibly non-orientable) surface with boundary represented as the union of two sets of discs, a set $V(G)$ of *vertices*, and a set $E(G)$ of *edges* such that

1. The vertices and edges intersect in disjoint line segments
2. Each such line segment lies on the boundary of precisely one vertex and precisely one edge
3. Every edge contains exactly two such line segments

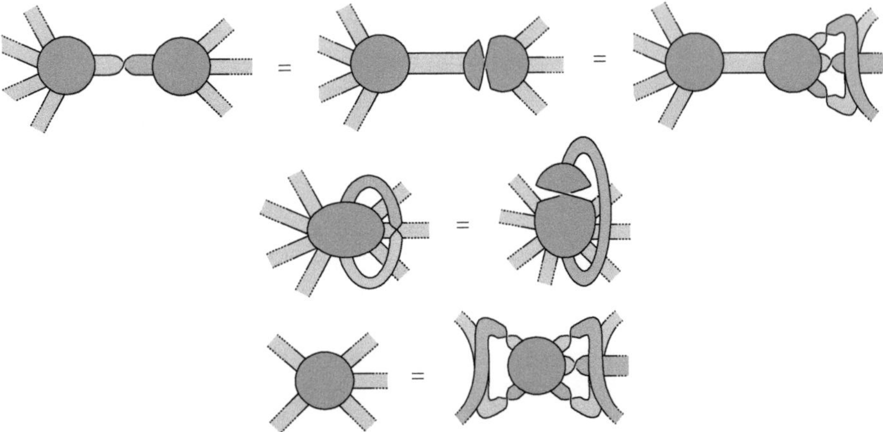

Fig. 1.4 Moving twists around a drawing of a ribbon graph

Figure 1.3c provides an example of a ribbon graph.

Ribbon graphs are known to be equivalent to cellularly embedded graphs. If $G \subseteq \Sigma$ is a cellularly embedded graph, a ribbon graph representation results from taking a small neighbourhood of G in Σ. On the other hand, if G is a ribbon graph, then it is a punctured surface. Capping off the punctures (i.e., sewing discs into each boundary component of the ribbon graph) results in a ribbon graph embedded in a closed surface. Contracting the ribbon graph in the surface then yields a graph embedded in the surface. This equivalence is illustrated in Fig. 1.3a–c.

Two ribbon graphs are *equivalent* if they define equivalent cellularly embedded graphs. Informally, this means that two ribbon graphs are equivalent if there is a homeomorphism taking one to the other that preserves the graph structure. We consider ribbon graphs up to this equivalence. In particular, we note that when we draw ribbon graphs, their embeddings in 3-space are irrelevant. In drawing ribbon graphs, we can "push" half-twists of edges around the ribbon graph and "turn vertices over" as illustrated in Fig. 1.4.

1.1.5 Band Decompositions

When passing from cellularly embedded graphs to ribbon graphs as described above, there is an intermediate stage, shown in Fig. 1.3b, which consists of a ribbon graph with its faces filled in with discs. This object is known as a *band decomposition*. The faces are called 2-*bands*, the edges are called 1-*bands*, and the vertices are called 0-*bands*.

Thus a band decomposition is obtained from a ribbon graph by capping off the punctures, and a ribbon graph is obtained from a band decomposition by

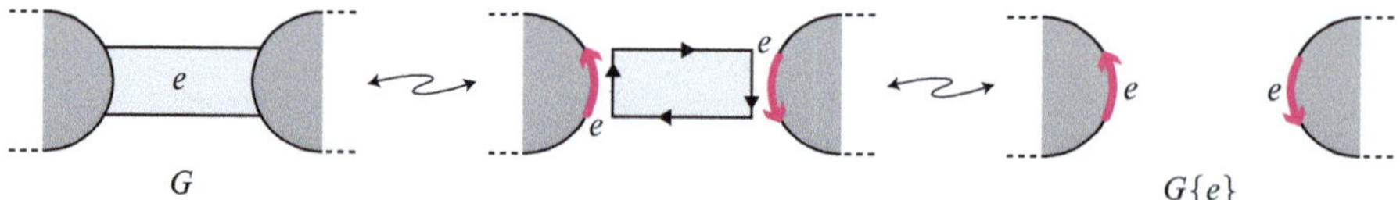

Fig. 1.5 Constructing $G\{e\}$ from a ribbon graph G

deleting the 2-bands. We can move between cellularly embedded graphs and band decompositions in one direction by thickening the cellularly embedded graph and in the other direction by contracting the 0-bands to points and the 1-bands to lines.

An advantage of band decompositions is that they may be drawn with equal visual weight given to faces and vertices, thus revealing some duality structures.

We refer the reader to [41] for further details on band decompositions.

1.1.6 Ribbon and Arrow Marked Graphs (Ram Graphs)

There is a natural way to encode the position of an edge e in a ribbon graph by a pair of e-labelled arrows on the boundary of vertices. Given a ribbon graph G and an edge $e \in E(G)$, arbitrarily orient the boundary of e; place an arrow on each of the two arcs where e meets vertices of G, such that the directions of these arrows follow the orientation of the boundary of e; label the two arrows with e; and delete the edge e. This process is illustrated locally at an edge in Fig. 1.5. This results in a ribbon graph, with exactly two e-labelled arrows on its boundary.

Conversely, given a ribbon graph with two e-labelled arrows on boundary arcs that meet vertices, we can recover a ribbon graph as follows: take a disc and orient its boundary arbitrarily; add this disc to the ribbon graph by choosing two non-intersecting arcs on its boundary and then identifying the arcs with the two e-labelled marking arrows matching the orientations of the boundary arcs with the directions of the marking arrows. The disc that has been added forms an edge of a new ribbon graph. Again, this process is illustrated in Fig. 1.5, only now reading from right to left.

Motivated by this way to encode the edges of a ribbon graph by pairs of arrows, we make the following definition.

Definition 1.6. A *ribbon and arrow marked graph*, or, more briefly, *ram graph*, consists of a ribbon graph equipped with a collection of labelled arrows, called *marking arrows*, on the intersection of its boundary components and vertices. The marking arrows have the property that no marking arrow meets an edge of the ribbon graph, and there are exactly two marking arrows with each label. See Fig. 1.3d, e.

If G is a ribbon graph and $A \subseteq E(G)$, then we can obtain a ram graph by replacing each edge in A with a pair of arrows as described above and in Fig. 1.5. Conversely, if G is a ram graph, then we can obtain a unique ribbon graph by replacing the pairs of arrows with edges again as described above and in Fig. 1.5. Thus every ram graph represents a unique ribbon graph.

These observations allow us to establish the following notation for ram graphs. We use $G\{A\}$ to denote a ram graph. Here A is the set of labels of the arrows, and G is the unique ribbon graph *represented by* the ram graph. Thus, if G is a ribbon graph, then $G\{A\}$ is the ram graph obtained by replacing the edges in $A \subseteq E(G)$ with pairs of arrows. See Fig. 1.3c–e.

Two ram graphs are *equivalent* if they represent the same ribbon graph. That is $G\{A\} = H\{B\}$ if and only if $G\{A\}$ and $H\{B\}$ represent the same ribbon graph if and only if $G = H$. (Note that A need not equal B.)

We emphasise the following two facts about ram graphs. Firstly, if $G\{A\}$ is a ram graph, then G is the ribbon graph it represents. G is *not* the underlying ribbon graph on which the arrows lie (unless $A = \emptyset$). Secondly, ram graphs are distinct from the arrow marked ribbon graphs introduced in [78]. This is because, although ram graphs and arrow marked ribbon graphs have the same underlying combinatorial objects, they are considered up to a different equivalence relation.

1.1.7 Arrow Presentations

Arrow presentations give a particularly efficient way to represent ribbon graphs. Every ribbon graph G has a representation as $G\{E(G)\}$ which consists of a collection of vertices with pairs of marking arrows on their boundary components. In such cases, to describe G it is enough to record only the marked boundary components of the vertex set (to recover the vertex set, just identify each of these boundary components with the boundary of a disc). Thus a ribbon graph can be presented as a set of circles with marking arrows on them. In such a structure, there are exactly two marking arrows with each label. Such a structure is called an arrow presentation. Formally:

Definition 1.7 (Chmutov [16]). An *arrow presentation* consists of a set of circles with pairs of labelled arrows, called *marking arrows*, on them such that there are exactly two marking arrows of each label.

An example of a ribbon graph and its representation as an arrow presentation is given in Fig. 1.3c, f. Arrow presentations are *equivalent* if they describe the same ribbon graph and are considered up to this equivalence.

We emphasise that the circles in an arrow presentation are not equipped with any embedding in the plane (or $\mathbb{R}^3$). Thus, in drawings, the circles may be distorted or flipped over, as in Fig. 2.5b, or they may have self-intersections as in Fig. 2.6c. Also, because of the equivalence relation on arrow presentations, reversing the direction of both arrows with a given label yields an equivalent arrow presentation.

1.1.8 Signed Rotation Systems

A *signed rotation system* consists of an abstract graph G, a cyclic ordering of the half-edges at each vertex, and a $+$ or $-$ sign on each edge.

Two signed rotation systems are *equivalent* if one can be transformed into the other by a sequence of local switches, where a local switch at a vertex means reversing the cyclic order of the half-edges about the vertex and toggling the sign on each of its incident edges (in the case of a loop at the vertex, the net is no change of sign on the loop edge). We consider signed rotation systems up to this equivalence. See [41, 81] for further details of signed rotation systems.

Signed rotation systems are equivalent to ribbon graphs. Given a ribbon graph, choose an orientation for each vertex disc and each edge. Use the orientation at each vertex disc to determine a rotation system for the underlying abstract graph. If the orientations on an edge agree with the orientations on both of its incident vertices, or disagree with both, assign a $+$ to the edge in the abstract graph. Otherwise assign a $-$. This gives a signed rotation system. The fact that signed rotation systems are equivalent under local switches means that this construction is independent of the choice of orientations on the vertex discs and edges of the ribbon graph. See Fig. 1.3g, h.

Conversely, given a signed rotation system, a ribbon graph results from creating vertex discs, giving each disc an orientation, and attaching the edge ribbons to these discs in the order given by the cyclic order at that vertex in the signed rotation system. Each edge ribbon is given an orientation, and an edge ribbon is attached with both ends consistent with the orientations of its incident vertices if its sign is $+$, and exactly one end inconsistent otherwise.

Note that although unsigned rotation systems appear in the literature associated with graphs embedded in orientable surfaces, we will not use them in that sense here. In the context of signed rotation systems, an embedded graph is orientable if it can be represented by a signed rotation system with a $+$ on all of its edges.

1.1.9 A Note on Terminology

We have seen that each of

- A cellularly embedded graph
- A ribbon graph
- A band decomposition
- A ram graph
- An arrow presentation
- A signed rotation system

can be used to describe the same object. We will use these equivalent representations interchangeably, using whichever best facilitates the discussion at hand. For simplicity, we use the term "embedded graph" loosely to mean any of the above equivalent representations of graphs on surfaces; in particular, we identify all representations of a given embedded graph. If G and H are equivalent embedded graphs in any of their realisations, we write $G = H$. We consider embedded graphs up to this equivalence. We caution though that our use of the term embedded graph is not consistent with some uses in the literature, where it can mean a graph drawn in a surface without the requirement that the components of $\Sigma \backslash G$ be discs.

In all representations, if G is an embedded graph, we let $f(G)$ denote the number of faces it has when represented as a cellularly embedded graph (this is equal to the number of boundary components for ribbon graphs), and we define $e(G)$, $v(G)$, $k(G)$, $r(G)$, and $n(G)$ to be the number of edges, vertices, components, rank, and nullity of the underlying abstract graph G. Finally, we let $\mathscr{G}$ denote the set of all embedded graphs and $\mathscr{G}_n$ the set of embedded graphs with exactly n edges.

1.2 Further Properties of Embedded Graphs

1.2.1 Subgraphs of Embedded Graphs

A technical consideration when working with a cellularly embedded graph $G \subset \Sigma$ is that removing one or more of its edges may result in a graph that is then not cellularly embedded in Σ. However, with a ribbon graph, removing vertices or edges results in another ribbon graph and therefore an embedded graph. This closure property of the set of ribbon graphs is one of our primary reasons for using them here.

A ribbon graph $H = (V(H), E(H))$ is a *ribbon subgraph* of $G = (V(G), E(G))$ if H can be obtained by deleting vertices and edges of G. If $V(H) = V(G)$, then H is a *spanning ribbon subgraph* of G. If $A \subseteq E(G)$, then the ribbon subgraph *induced* by A, denoted $G|_A$, is the ribbon subgraph of G that consists of the edges in A and their incident vertices. We will often regard H as being embedded in G and will often identify the vertices and edges of H with the corresponding vertices and edges of G.

We define embedded subgraphs for other realisations of embedded graphs by translating into the language of ribbon graphs. For example, a cellularly embedded graph $H \subset \Sigma'$ is a *cellularly embedded subgraph* of $G \subset \Sigma$ if, when translating to the language of ribbon graphs, H is a ribbon subgraph of G. Note that Σ' need not equal Σ. Ram subgraphs, sub-rotation systems, sub-band decompositions, and sub-arrow presentation can be defined similarly.

Just as with abstract graphs, if G is a ribbon graph and $A \subseteq E(G)$, we let $e(A)$, $v(A)$, $k(A)$, $r(A)$, and $n(A)$ denote the number of edges, vertices, components, rank, and nullity of the spanning ribbon subgraph $(V(G), A)$ of G.

1.2.2   Genus and Loops

A cellularly embedded graph $G \subset \Sigma$ is said to be *orientable* if each component of Σ is orientable; otherwise G is said to be *non-orientable*. If G is connected, then the *genus*, $g(G)$, of G is the genus of Σ, otherwise its genus is the sum of the genera of its components. A cellularly embedded graph $G \subset \Sigma$ is a *plane graph* if Σ is the 2-sphere, S^2.

Translating into the language of ribbon graphs gives that a ribbon graph is *orientable* if it is orientable when viewed as a surface (here we are again using the fact that a ribbon graph is a punctured surface). Similarly, the genus, $g(G)$, of a ribbon graph G is its genus when viewed as a punctured surface. A ribbon graph G is a *plane ribbon graph* if it is connected and $g(G) = 0$.

The *Euler genus* $\gamma(G)$ of a ribbon graph or cellularly embedded graph G is defined as follows. If G is a connected, then

$$\gamma(G) := \begin{cases} 2g(G), & \text{if } G \text{ is orientable;} \\ g(G), & \text{if } G \text{ is non-orientable.} \end{cases}$$

If G is not connected, then $\gamma(G)$ is defined as the sum of the value of γ of each of its components.

Orientability, genus, and Euler genus are defined for band decompositions, signed rotation systems, ram graphs, and arrow presentations, and other realisations of embedded graphs by translating to the language of cellularly embedded graphs or ribbon graphs.

The *Euler characteristic*, $\chi(G)$, of an embedded graph G, is defined by

$$\chi(G) = v(G) - e(G) + f(G).$$

The Euler characteristic can be shown to be related to the Euler genus by the following formula:

$$\chi(G) = v(G) - e(G) + f(G) = 2k(G) - \gamma(G).$$

A loop (respectively, cycle) in a cellularly embedded graph is said to be a *non-orientable loop* (respectively, *non-orientable cycle*) if a neighbourhood of it (including the vertex incident to it, in the case of a loop) is homeomorphic to a Möbius band. Otherwise it is said to be an *orientable loop* (respectively, *orientable cycle*). Orientable and non-orientable loops and cycles are defined for other realisations of embedded graphs by translating into the language of cellularly embedded graphs.

As an example, in Fig. 1.6a, the two loops 1 and 6 are non-orientable, while the two loops 2 and 7 are orientable. Also, the two cycles given by $\{3,4\}$ and $\{3,5\}$ are non-orientable, while that given by $\{4,5\}$ is orientable.

Let C and C' be two cycles in an embedded graph G. Further suppose v is a vertex in both cycles, with edges $e, f \in C$ and edges $e', f' \in C'$ all incident with v (possibly

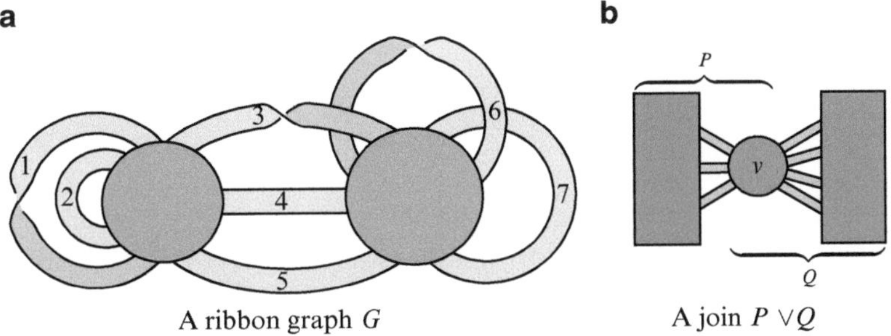

A ribbon graph G A join $P \vee Q$

Fig. 1.6 An illustration of the definitions in Sect. 1.2.2

$e = f$ or $e' = f'$ if C or C' is a 1-cycle). We then say that C and C' *alternate* at v if there is an orientation of a neighbourhood of v so that the induced cycling ordering of the edges incident with v has e and f alternating with e' and f', that is, they appear in the cyclic order as $(e \cdots e' \cdots f \cdots f' \cdots)$. We say that two loops e and f alternate if their induced subgraphs are alternating 1-cycles. Similarly, a loop e and a cycle C alternate if the 1-cycle induced by e alternates with C. For example, in Fig. 1.6a, the loops 6 and 7 alternate, while the loops 1 and 2 do not. The loop 6 and the cycle given by $\{3,4\}$ alternate, but the loop 6 and the cycle given by $\{4,5\}$ do not.

A loop e at a vertex v of an embedded graph G is *trivial* if there is no cycle or edge in G which alternates with e. In Fig. 1.6a, the loops 1 and 2 are trivial, but the loops 6 and 7 are non-trivial.

If G, P, and Q are ribbon graphs, then we say that G is the *join* of P and Q, written $G = P \vee Q$, if G can be obtained by identifying an arc on a vertex of P with an arc on a vertex of Q as indicated in Fig. 1.6b. The two arcs that are identified should not intersect any edges. Joins are defined for embedded graphs generally, by translating into the language of ribbon graphs. The join is also known as the "one-point join", the "map amalgamation", and the "connected sum" in the literature. We emphasise that if C and C' are alternating cycles or loops in $P \vee Q$, then C and C' are alternating cycles or loops in P or in Q. For example, referring to Fig. 1.6a, if P is the subgraph induced by $\{1,2\}$ and Q is the subgraph induced by $\{3,4,5,6,7\}$, then $G = P \vee Q$. However, if P' is the subgraph induced by $\{6,7\}$ and Q' is the subgraph induced by $\{1,2,3,4,5\}$, then $G \neq P' \vee Q'$.

We say that G is the *disjoint union* of P and Q, written $G = P \sqcup Q$, if $G = P \cup Q$ and $P \cap Q = \emptyset$.

The Euler genus is additive under joins and disjoint unions:

$$\gamma(P \sqcup Q) = \gamma(P \vee Q) = \gamma(P) + \gamma(Q).$$

Note, however, that in general the genus g is not additive under joins.

1.3 Petrials of Embedded Graphs

Two operations on embedded graphs, forming the Petrie dual and the geometric dual, are the foundation of the ribbon group action which we define in Chap. 2 and from which we derive most of the results in the subsequent chapters. Petrie duals, also called Petrials, are particularly easy to define in terms of both ribbon graphs and arrow presentations, so we begin with them rather than the possibly more familiar geometric duals.

The *Petrial* of a ribbon graph G, denoted $G^{\times}$, is formed by detaching one end of each edge from its incident vertex disc, giving the edge a half-twist, and reattaching it to the vertex disc as shown in Fig. 1.7. An example of a ribbon graph and its Petrial is given in Fig. 1.8.

The Petrial of an arrow presentation is formed by reversing the direction of exactly one of the arrows for each label. For a ram graph, the ribbon edges are each given a half-twist, while the arrow marked edges each have one arrow reversed. For signed rotation systems the Petrial is obtained by toggling the sign of each edge. We define Petrials for cellularly embedded graphs by translating into the language of ribbon graphs or arrow presentations. We note that the Petrial, $G^{\times}$, is the embedded graph with the same edges and vertices as G, but with Petrie polygons, which are the result of closed left-right walks in G, as the faces (see Wilson [100]).

1.4 Geometric Duality

Geometric duality is a fundamental construction in graph theory and appears in many areas of mathematics. Let $G \subset \Sigma$ be a cellularly embedded graph. The *geometric dual* $G^{*} \subset \Sigma$ is the cellularly embedded graph obtained from G by placing one vertex in each of its faces and embedding an edge of G^{*} between two

Fig. 1.7 Giving a half-twist to an edge of a ribbon graph

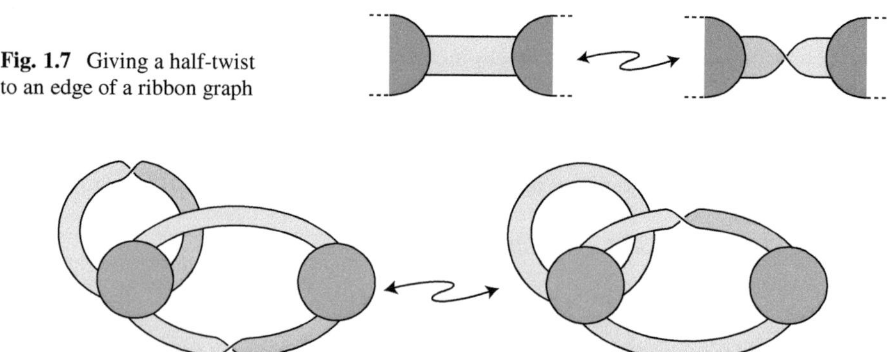

Fig. 1.8 A ribbon graph G and its Petrial $G^{\times}$

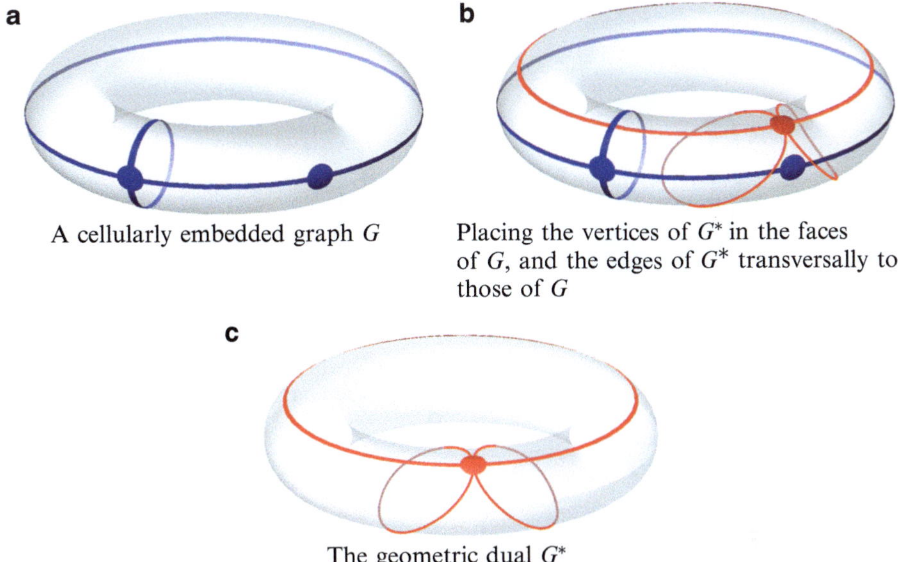

a A cellularly embedded graph G

b Placing the vertices of G^* in the faces of G, and the edges of G^* transversally to those of G

c The geometric dual G^*

Fig. 1.9 Forming a geometric dual of a cellularly embedded graph

of these vertices whenever the faces of G that they lie in are adjacent. The edges of G^* are embedded so that they cross the corresponding face boundary (or edge of G) transversally.

Figure 1.9 illustrates the construction of the geometric dual of a graph cellularly embedded in the torus.

There is a natural bijection between the edges of G and the edges of G^*, and we use it to identify the edges of G and G^*. Observe that duality acts disjointly on the components of a cellularly embedded graph. Proposition 1.8 follows directly from the construction of geometric duals.

Proposition 1.8. *Let G be an embedded graph. Then*

1. $(G^*)^* = G$
2. $\gamma(G) = \gamma(G^*)$

Geometric duals are particularly easy to construct in the language of ribbon graphs. We sew discs, which become the vertices $V(G^*)$ of G^*, into the boundary components of G (i.e., the discs cap-off the punctures). The geometric dual G^* then results when the original vertex discs of G are removed. See Fig. 1.10.

Constructing G^* is even easier in terms of band decompositions. The geometric dual G^* results from simply reversing the roles of the 0-bands (the vertices) and the 2-bands (the faces) of G.

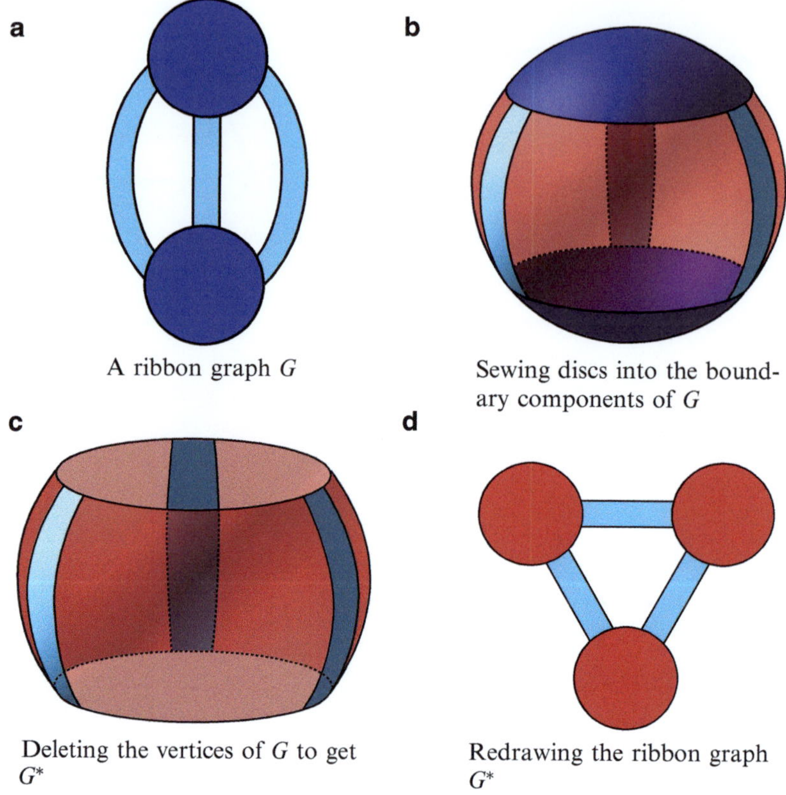

A ribbon graph G

Sewing discs into the boundary components of G

Deleting the vertices of G to get G^*

Redrawing the ribbon graph G^*

Fig. 1.10 Forming the geometric dual of a ribbon graph

Geometric duals of ram graphs, arrow presentations, and signed rotation systems may be defined by translating to the language of cellularly embedded graphs or ribbon graphs. Note that for ram graphs the identity $(G\{A\})^* = G^*\{A\}$ holds.

1.5 Medial Graphs, Tait Graphs, and Duality

We now turn to two other essential constructions: those of medial graphs and Tait graphs. These graphs are not only important in themselves, for example by enabling interpretations and constructions of graph polynomials as in Chap. 4, but they also provide a bridge between knot theory and graph theory, which is the topic of Chap. 5. We review these constructions in some detail, since in Chap. 2 we will fully generalise duality to graphs on surfaces, which will lead in turn to a full generalisation of Tait graphs via medial graphs, and these play a central role in the subsequent theory.

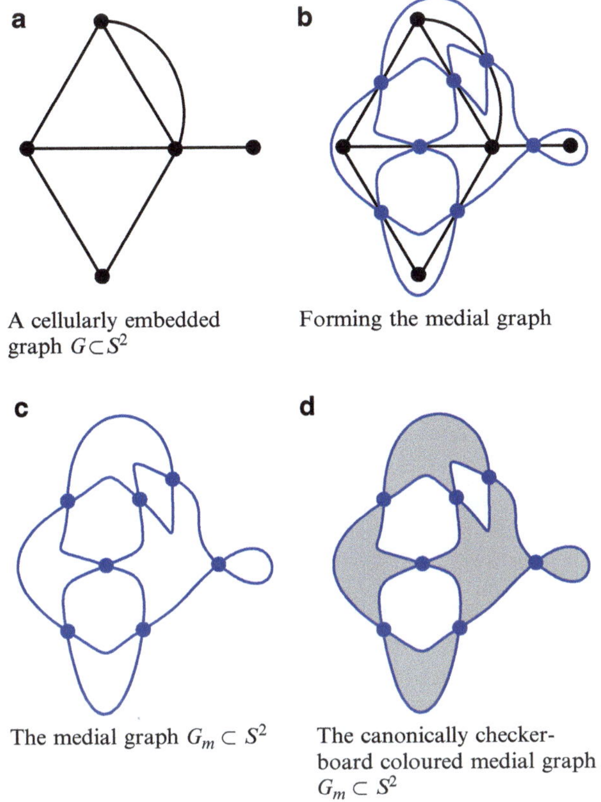

a A cellularly embedded graph $G \subset S^2$

b Forming the medial graph

c The medial graph $G_m \subset S^2$

d The canonically checkerboard coloured medial graph $G_m \subset S^2$

Fig. 1.11 Forming a medial graph

1.5.1 Medial Graphs

If a graph G is cellularly embedded, we construct its *medial graph* G_m by placing a vertex of degree 4 on each edge of G and then drawing the edges of the medial graph by following the face boundaries of G. See Fig. 1.11a–c. Consistent with this definition is that the medial graph of an isolated vertex is an isolated face, and we adopt this convention. Medial graphs may thus have "free loops", where a free loop is a circular edge with no incident vertex. Note that if G is cellularly embedded in a surface Σ, then G_m is also cellularly embedded in Σ.

The medial graphs of ribbon graphs, arrow presentations, signed rotation systems, and ram graphs are most easily formed by translating to the language of cellularly embedded graphs, forming the medial graphs, and translating back.

A *checkerboard colouring* of a cellularly embedded graph is an assignment of the colour black or white to each face such that adjacent faces receive different colours (i.e., it is a face 2-colouring). A medial graph is always 4-regular and, if viewed as cellularly embedded, can always be checkerboard coloured by colouring the faces

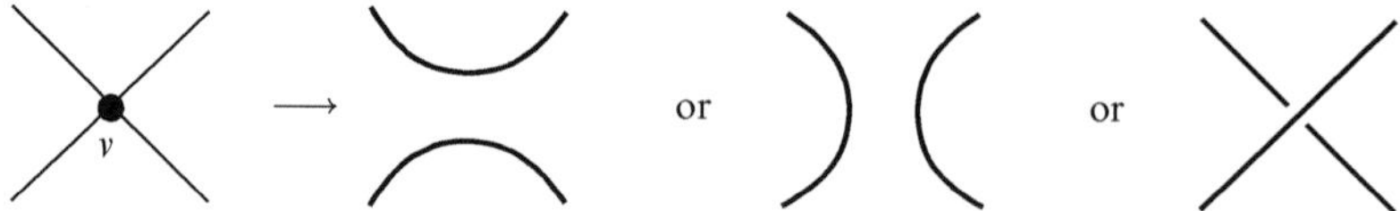

Fig. 1.12 The three vertex states of a vertex v of an abstract graph

containing a vertex of the original graph black and the remaining faces white. We call this checkerboard colouring of a medial graph G_m the *canonical checkerboard colouring*. Note that G must be specified to determine the canonical checkerboard colouring of G_m. Figure 1.11d shows the canonical checkerboard colouring of the medial graph G_m of Fig. 1.11c. While not all 4-regular embedded graphs admit checkerboard colourings, we have seen that all medial graphs do. In addition all 4-regular plane graphs do:

Proposition 1.9. *Let F be a plane graph.*

1. *Then F is checkerboard colourable if and only if it is even (i.e. all vertices have even degree).*
2. *If F is 4-regular, then it is checkerboard colourable.*

Proof. If F is checkerboard colourable, then clearly F is even, since the face colours alternate about each vertex.

Now assume F is even and is drawn on the plane. Note that F can be decomposed into a set of edge disjoint cycles by removing any cycle and applying induction. These cycles form a collection of simple closed curves in the plane, so it suffices to prove that the regions formed by any such collection may be checkerboard coloured. We proceed by induction on the number of curves. It is true for one curve since, by the Jordan Curve Theorem, this curve has an inside and an outside, which may be given opposite colours. Now assume there are $n-1$ simple closed curves in the plane with checkerboard coloured regions and add an nth such curve. Again, by the Jordan Curve Theorem, this curve has an inside and an outside, and the desired checkerboard colouring is achieved by switching the colours on the regions inside the curve.

The second item in the proposition follows from the first item.   □

1.5.2 Vertex States and Graph States

Vertex states and graph states play a central role in many of the constructions in this text. A vertex state at a vertex v of a 4-regular graph arises from a partition, into pairs, of the half-edges incident to v. If (u,v) and (w,v) are two non-loop edges whose half edges are paired at the vertex v, then we replace these two edges with a single edge (u,w), which, in the case of a cellularly embedded graph, lies in an ε-neighbourhood of the original two edges. In the case of a loop at the vertex, we temporarily imagine an extra vertex of degree two on the loop, carry out the operation, and then remove the temporary vertex. After this is done for all pairs in the partition at v, the resulting configuration is called a *vertex state* at v. See

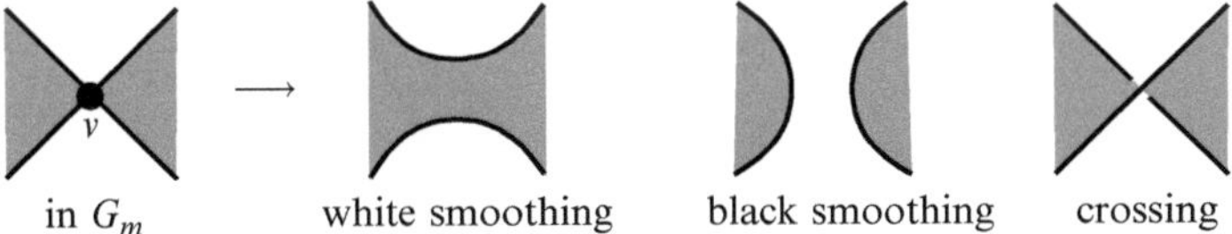

Fig. 1.13 The three vertex states of a vertex v of a checkerboard coloured embedded graph

Fig. 1.12. For general abstract graphs, there is no way to distinguish among the three vertex states.

However, if G is an embedded graph and G_m its canonically checkerboard coloured medial graph, then we may use the checkerboard colouring to distinguish among the vertex states at v as in Fig. 1.13. We call the three vertex states a *white smoothing*, a *black smoothing*, and a *crossing* as defined in the figure. The white smoothing and black smoothing are known collectively as *smoothings*.

This kind of operation at a vertex appears in the literature under several names, including "split", "smoothing", and "transition". The operations were referred to as "splittings" in [30]. However, because we will be concerned with knots in Chap. 5, we will here use the term "smoothing" as this is more prevalent in current knot theory literature.

A *graph state s* of any 4-regular graph F is a choice of vertex state at each of its vertices. A graph state s consists of a set of disjoint closed curves, and we refer to these as the *components of the state*, denoting the number of them by $c(s)$.

It will be useful to have some notation for the embedded graphs obtained by replacing a vertex v with one of its vertex states.

Notation 1.10. If G_m is a canonically checkerboard coloured medial graph, then we denote the graphs that result from each of the vertex states at v as follows:

$(G_m)_{wh(v)}$ is the embedded graph that results from taking the white smoothing at the vertex v of G_m.

$(G_m)_{bl(v)}$ is the embedded graph that results from taking the black smoothing at v of G_m.

$(G_m)_{cr(v)}$ is the embedded graph that results from taking the crossing at v of G_m.

1.5.3 Tait Graphs

Suppose we are given an embedded 4-regular graph F. The question arises as to whether it is the medial graph of some embedded graph G, and if so, what are all the embedded graphs G with embedded medial graphs equal to F. This is readily answered, but we need a few definitions first.

Definition 1.11. Let F be a checkerboard coloured cellularly embedded graph. Then

1. The *blackface graph*, F_{bl}, of F is the embedded graph constructed by placing one vertex in each black face and adding an edge between two of these vertices whenever the corresponding regions meet at a vertex of F.

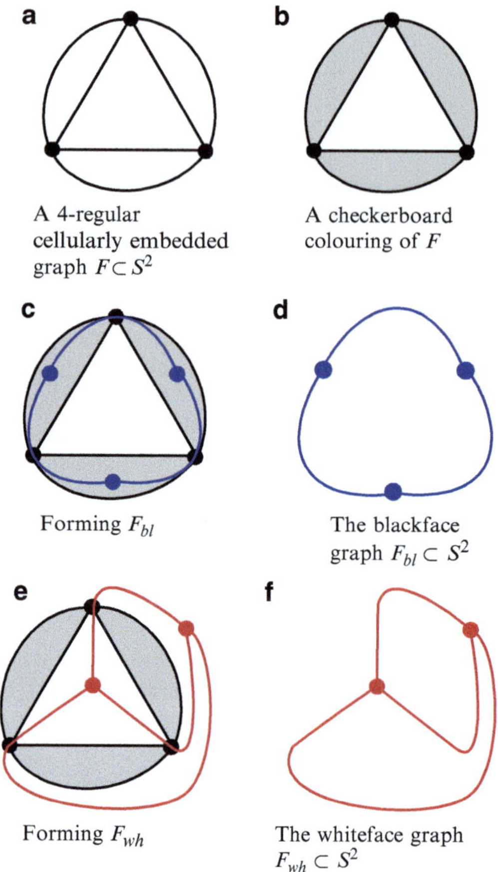

a A 4-regular cellularly embedded graph $F \subset S^2$

b A checkerboard colouring of F

c Forming F_{bl}

d The blackface graph $F_{bl} \subset S^2$

e Forming F_{wh}

f The whiteface graph $F_{wh} \subset S^2$

Fig. 1.14 Forming Tait graphs

2. The *whiteface graph*, F_{wh}, is constructed analogously by placing vertices in the white faces.

The constructions of the whiteface and blackface graphs are illustrated in Fig. 1.14. The whiteface and blackface graphs are defined for other presentations of embedded graphs by translating into the language of cellularly embedded graphs.

Note that if F is checkerboard coloured, then switching the colours of all of the faces results in a second checkerboard colouring of F, so that F_{wh} (respectively, F_{bl}) for one of these colourings is equal to F_{bl} (respectively, F_{wh}) for the other.

If F is a checkerboard colourable but not checkerboard coloured, we can assign a checkerboard colouring to F and form the whiteface and blackface graphs with respect to this checkerboard colouring. Since F was originally not checkerboard coloured, there is no consistent assignment of which graph is the blackface graph and which is the whiteface graph; accordingly we refer to both of graphs as Tait graphs of F, a term borrowed from knot theory, as follows.

Definition 1.12. Let F be a checkerboard colourable embedded graph. Then the two *Tait graphs* are the embedded graphs obtained by checkerboard colouring F and forming the whiteface and blackface graphs. We let $\mathscr{T}(F)$ denote the set of Tait graphs of F.

Remark 1.13. Historically Tait graphs were considered for 4-regular graphs cellularly embedded in the plane $\mathbb{R}^2$. In this case we can canonically checkerboard colour the graph by insisting that the unbounded region be white. In the literature, the term Tait graph is often used to refer to the (unique) blackface graph of F formed with respect to this canonical checkerboard colouring. Here, however, we consider graphs embedded in closed surfaces (in particular, our plane graphs are drawn on S^2 rather than $\mathbb{R}^2$) and so there is no canonical checkerboard colouring. Accordingly, here we cannot define a unique Tait graph, only a pair of Tait graphs.

The following well-known results provide the principal motivation for Chaps. 2 and 3.

Theorem 1.14. *Let F be a 4-regular embedded graph. Then F is the medial graph of some embedded graph G if and only if F is checkerboard colourable. If F is checkerboard colourable, then F is the embedded medial graph of precisely F_{bl} and F_{wh}.*

Proof. First suppose that $F = G_m \subset \Sigma$. In the construction of G_m from G there is a canonical immersion of $G \cup G_m$ in Σ, which is illustrated in Fig. 1.11b, in which each vertex of G_m lies on a unique edge of G and the edges of G_m follow the faces of G. We can obtain a checkerboard colouring of G_m by colouring each face of G_m that contains a vertex of G black (which is the canonical checkerboard colouring) or by colouring each face of G_m that does not contain a vertex of G black. These are all of the checkerboard colourings of G_m since an embedded graph can have at most two checkerboard colourings. If the faces of G_m that contain a vertex of G are coloured black, it is readily seen that $G = (G_m)_{bl} = F_{bl}$; and if the faces of G_m that do not contain a vertex of G are coloured black, it is readily seen that $G = (G_m)_{wh} = F_{wh}$.

Conversely, suppose that F is checkerboard colourable. Give F a checkerboard colouring. Then we can form F_{bl} and F_{wh}. By comparing the constructions of medial graphs and Tait graphs, it is easily seen that $(F_{bl})_m = F$ and $(F_{wh})_m = F$, as required. $\qquad\square$

Theorem 1.15. *Let G be an embedded graph and F be a 4-regular checkerboard coloured embedded graph. Then*

1. $\{F_{bl}, F_{wh}\} = \{G \mid G_m = F\}$, *or, equivalently,* $G_m = F \iff G = F_{bl}$ *or* $G = F_{wh}$
2. $(F_{bl})_m = (F_{wh})_m = F$
3. $\{(G_m)_{bl}, (G_m)_{wh}\} = \{G, G^*\}$
4. $F_{bl} = (F_{wh})^*$
5. $\{G, G^*\} = \{H \mid H_m = G_m\}$, *or, equivalently,* $H_m = G_m \iff H \in \{G, G^*\}$

Proof. The first item follows directly from Theorem 1.14. The second item follows from the proof of Theorem 1.14. The third item follows from the observation that if G_m is canonically checkerboard coloured, then the white faces of G_m correspond to the faces of G.

For the fourth item, suppose that $F_{bl} = G$. By Item 2, we have $(F_{bl})_m = G_m = F$. Thus $F_{bl} = (G_m)_{bl}$ and $F_{wh} = (G_m)_{wh}$. By Item 2 one of these two graphs must be G and the other must be G^*, and the result follows upon observing that $(G^*)^* = G$.

Finally, for the fifth item, Item 1 gives that $H_m = G_m$ if and only if $H = (G_m)_{bl}$ or $H = (G_m)_{wh}$. By Item 3 this happens if and only if $H = G$ or $H = G^*$. $\qquad\square$

Theorem 1.14 answers the original question of this section of when a 4-regular embedded graph is a medial graph and of what graphs. Theorem 1.15 shows that geometric duality, Tait graphs, and medial graphs are intimately connected. In Chap. 3 we will fully generalise Theorem 1.15 to graphs on surfaces with the more complete understanding of duality developed in Chap. 2. The second half of Chap. 3 will show the effects on the two sides of Item 5 of Theorem 1.15 under a hierarchy of equivalences and dualities.

Chapter 2
Generalised Dualities

This chapter introduces the framework upon which we will build the rest of this monograph. We begin by considering geometric duals and Petrials, observing that both Petriality and geometric duality result from local operations on each edge of an embedded graph. These local operations applied to subsets of the edge set result in partial Petriality and partial duality. We provide constructions for partial duals and partial Petrials in various realisations of embedded graphs. The two operations of partial Petriality and partial duality give rise to an action of the symmetric group on embedded graphs with a distinguished set of edges. This group action leads to twisted duality, which assimilates several types of duality from the literature, including geometric duality, direct derivatives, Petrie duals, and partial duality. We concluded by defining the ribbon group and describing how twisted duality can be obtained as orbits under the ribbon group action on the set of edge-ordered embedded graphs.

2.1 Partial Petrials

Recall from Sect. 1.3 that if G is a ribbon graph, then its Petrial, $G^{\times}$, is obtained by adding a half-twist to each edge of G, as in Fig. 1.7. Notice that rather than adding a half-twist to every edge of G, we may add half-twists to only some of the edges of G. The result is a partial Petrial:

Definition 2.1. Let G be a ribbon graph and $A \subseteq E(G)$. Then the *partial Petrial*, $G^{\tau(A)}$, of G with respect to A is the ribbon graph obtained from G by adding a half-twist to each of the edges in A, as in Fig. 1.7.

An example of a partial Petrial is shown in Fig. 2.1.

If $A = \{e\}$, we write $G^{\tau(e)}$ for $G^{\tau(\{e\})}$. Also, in parallel with the terminology for duals and duality, we can regard the act of taking a partial Petrial as an operation on embedded graphs which we call partial Petriality.

J.A. Ellis-Monaghan and I. Moffatt, *Graphs on Surfaces: Dualities, Polynomials, and Knots*, SpringerBriefs in Mathematics, DOI 10.1007/978-1-4614-6971-1_2, © Joanna A. Ellis-Monaghan, Iain Moffatt 2013

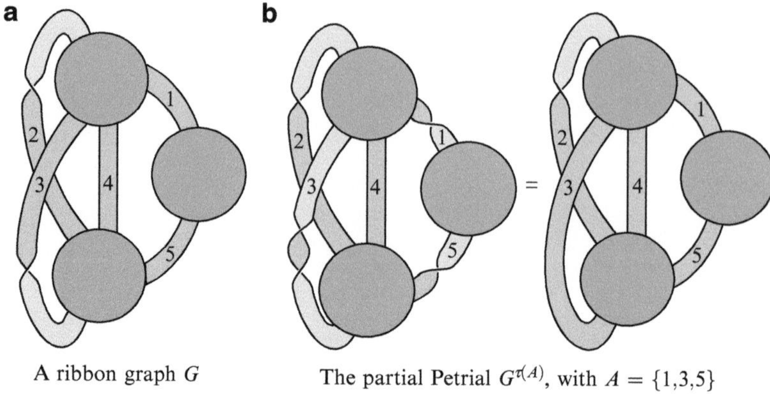

A ribbon graph G The partial Petrial $G^{\tau(A)}$, with $A = \{1,3,5\}$

Fig. 2.1 An example of a partial Petrial of a ribbon graph

Partial Petrials have a very simple construction in the setting of arrow presentations. If G is an arrow presentation and $A \subseteq E(G)$, then $G^{\tau(A)}$ is the arrow presentation obtained by, for each $e \in A$, reversing the direction of exactly one of the e-labelled arrows. In terms of signed rotation systems, $G^{\tau(A)}$ is formed by toggling the sign of every edge in A. Finally, partial Petrials for ram graphs and cellularly embedded graphs can be constructed by translating to, say, the language of ribbon graphs, taking the partial Petrial, and then translating back.

We observe a few basic properties of partial Petrials.

Proposition 2.2. *Let G be an embedded graph and $A, B \subseteq E(G)$.*

1. *$G^{\tau(\emptyset)} = G$.*
2. *$G^{\tau(E(G))}$ is the Petrial $G^\times$ of G.*
3. *$(G^{\tau(A)})^{\tau(B)} = G^{\tau(A\Delta B)}$, where $A\Delta B := (A \cup B) \setminus (A \cap B)$ is the symmetric difference of A and B.*
4. *Partial Petriality acts disjointly on components, i.e., $(P \sqcup Q)^{\tau(A)} = (P^{\tau(A \cap E(P))}) \sqcup (Q^{\tau(A \cap E(Q))})$.*
5. *$G^{\tau(\{e,f\})} = (G^{\tau(e)})^{\tau(f)} = (G^{\tau(f)})^{\tau(e)}$. In particular, partial Petrials can be formed one edge at a time.*
6. *There is a natural 1–1 correspondence between the edges of G and the edges of $G^{\tau(A)}$.*

2.2 Partial Duals

We have just seen that partial Petrials arise from the same local operation on the edges as generate Petrials, but by applying it to only a subset of edges of an embedded graph. In this section we show that the same principle applies to duality.

There is a local operation on individual edges that gives rise to geometric duals, and this operation can be applied to a subset of edges of an embedded graph, with a partial dual as the result.

Partial duality was introduced by Chmutov in [16] to relate various recent realisations of the Jones polynomial as evaluations of Bollobás and Riordan's ribbon graph polynomial (this will be discussed in detail in Chap. 5). Since its introduction, partial duality has found a variety of applications to knot theory [3, 13, 14, 16, 45, 77, 79], graph theory [16, 28–30, 44, 76, 78, 80, 96], and physics [42, 63, 89–91]. Partial duality appears to be a fundamental and far-reaching construction for embedded graphs.

As with partial Petrials, partial duals may be described in the various presentations for embedded graphs. Each description has its advantages and disadvantages: some properties can be immediately clear in one construction, but very hard to see in another. We will give descriptions for arrow presentations, ribbon graphs, and ram graphs. As with partial Petrials, partial duals for other expressions may be found by first translating to one of these three descriptions, finding the partial dual, and translating back.

2.2.1 Partial Duality with Respect to an Edge

To see that geometric duality is the result of a local operation applied to each edge of an embedded graph, we begin by drawing an embedded graph G as a band decomposition, giving equal emphasis to the vertices and faces, as in Fig. 2.2b. In Fig. 2.2b, the band decomposition for G has 0-bands A, B, and C; 1-bands 1, 2, 3, and 4; and 2-bands a, b, and c. Since a band decomposition for G^* can be obtained by switching the 0-bands and 2-bands, G^* has 0-bands a, b, and c; 1-bands 1, 2, 3, and 4; and 2-bands A, B, and C. We superimpose the circles of an arrow presentation of G on this drawing (as in Fig. 2.2c), and we superimpose the circles of an arrow presentation of G^* on another copy (as in Fig. 2.2d).

Consider a single edge e in the arrow presentation of both G and G^*. We see that the process of taking the dual is the result of a local operation on e that simply shifts the arrows associated with e from vertex boundaries to facial boundaries, as in Fig. 2.3.

Figure 2.3 therefore suggests an operation on arrow presentations that allows for the formation of the dual of an embedded graph with respect to a *single* edge: simply change the arrow presentation as indicated in the figure. This results in an arrow presentation for a ribbon graph $G^{\delta(e)}$ which is called the partial dual with respect to e. More generally, one may form the dual with respect to a subset of edges of G:

Definition 2.3. Let G be an arrow presentation and $A \subseteq E(G)$. Then the *partial dual*, $G^{\delta(A)}$, of G with respect to A is the arrow presentation obtained as follows. For each $e \in A$, suppose α and β are the two arrows labelled e in the arrow presentation of G. Draw a line segment with an arrow on it directed from the head

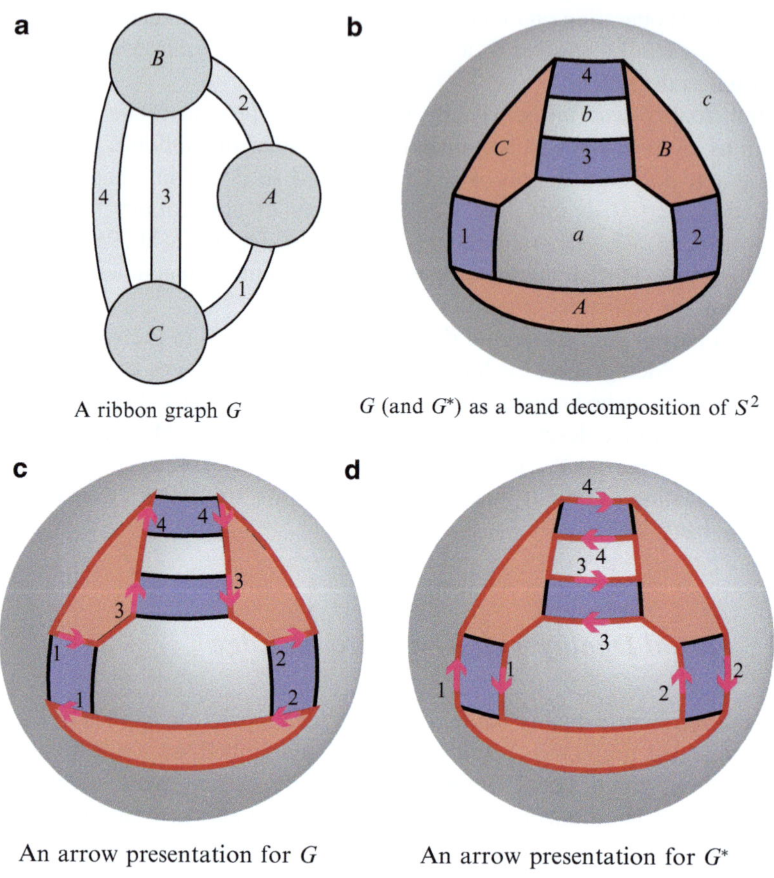

a A ribbon graph G

b G (and G^*) as a band decomposition of S^2

c An arrow presentation for G

d An arrow presentation for G^*

Fig. 2.2 An example of a partial dual

Fig. 2.3 Local dual operation at an edge e of G and G^*

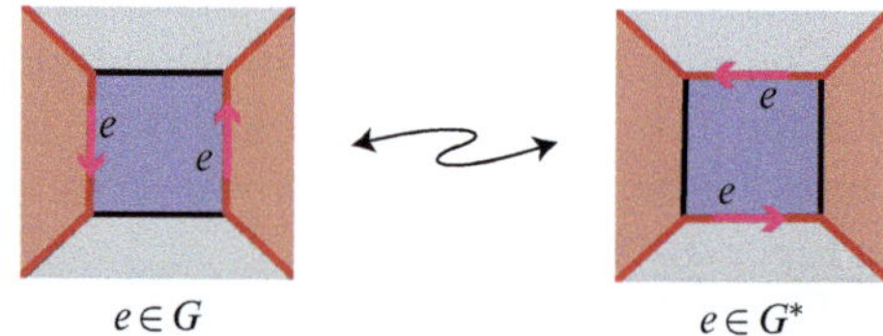

of α to the tail of β, and a line segment with an arrow on it directed from the head of β to the tail of α. Label both of these arrows e and delete α and β and the arcs containing them. This process is illustrated locally at a pair of arrows in Fig. 2.4.

Partial duals for cellularly embedded graphs, ribbon graphs, band decompositions, ram graphs, and signed rotation systems may be defined by translating to the

Fig. 2.4 Taking the partial dual of an edge in an arrow presentations

language of arrow presentations, forming the partial dual, and then translating back again. However, in the next section, we will describe some other ways to construct partial duals in these realisations of embedded graphs.

If $A = \{e\}$, we will often write $G^{\delta(e)}$ for $G^{\delta(\{e\})}$.

Figure 2.5 shows the partial dual $G^{\delta(3)}$ (Fig. 2.5b) of an arrow presentation G (Fig. 2.5a). The figure also shows the arrow presentations superimposed on the band decomposition of G and G^* (Fig. 2.5c, d) and also G and $G^{\delta(3)}$ as ribbon graphs (Fig. 2.5e, f). Observe that although both G and $G^{\delta(3)}$ are non-orientable, G is of genus 0, while $G^{\delta(3)}$ is of genus 1. Thus, unlike geometric duals, a partial dual may have a different genus than that of the original embedded graph.

Figure 2.6 gives a second example of a partial dual formed using Definition 2.3. In this figure the partial dual is formed with respect to $A = \{1,3\}$.

2.2.2 Other Constructions of Partial Duals

A number of methods, using different representations of embedded graphs, for constructing partial duals have appeared in the literature. Since each of these methods has its advantages, we review them all here and prove their equivalence. These methods typically use a more "global" approach, considering the entire subset of edges in the construction, as opposed to the "local", one edge at a time, approach we have just seen.

2.2.2.1 Partial Duals via Spanning Ribbon Subgraphs

We begin with Chmutov's original definition of partial duality from [16]. Let G be a ribbon graph and $A \subseteq E(G)$. Observe in Fig. 2.3 that in forming the partial dual using an arrow presentation, if $e \in A$, then we can position the arrow presentation so that it follows the boundaries of the edge e of G that coincide with a vertex as in the left-hand side of Fig. 2.3; and if $e \notin A$, then we can position the arrow presentation so that it follows the vertices incident to e of G that coincide with a face as in the right-hand side of Fig. 2.3. Thus we can position the arrow presentation for $G^{\delta(A)}$ so that it lies on the spanning ribbon subgraph $(V(G),A)$ of G. Moreover, the two e-labelled arrows in the arrow presentation all lie on the two arcs where the boundary of the edge e of G intersects the spanning ribbon subgraph. Thus, we may rewrite Definition 2.3 in the following way.

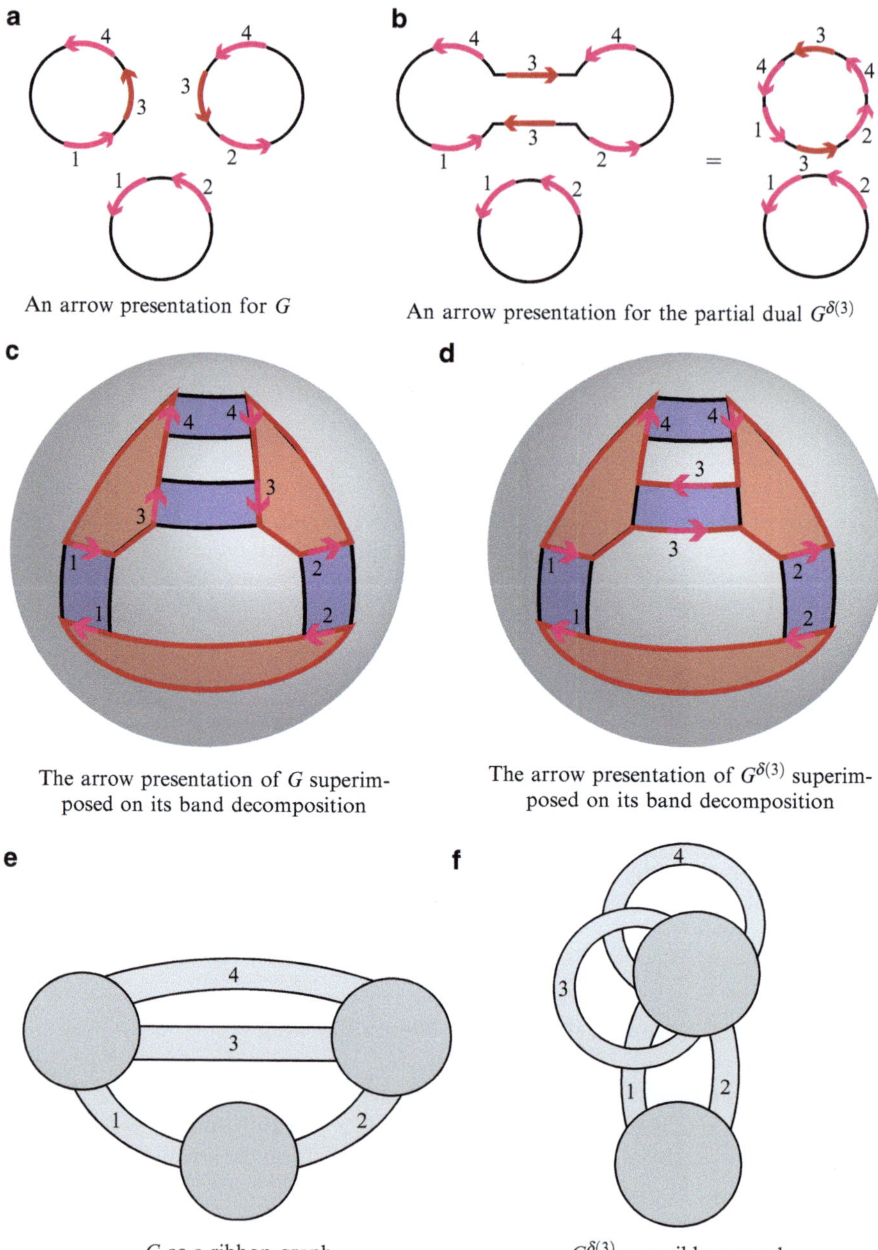

Fig. 2.5 Taking the partial dual with respect to the edge labelled 3

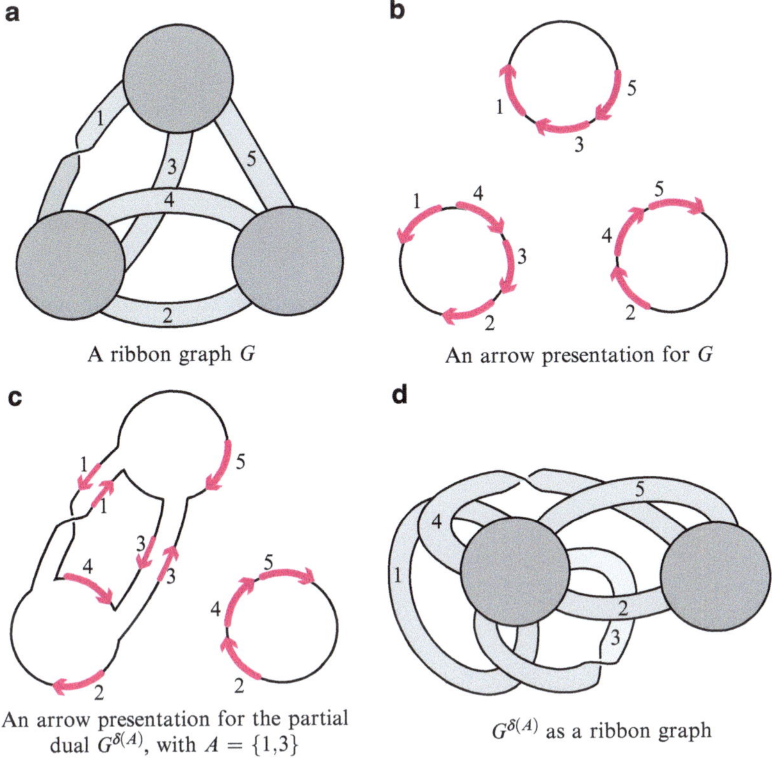

Fig. 2.6 An example of a partial dual formed using arrow presentations

Definition 2.4 (Chmutov [16]). Let G be a ribbon graph and $A \subseteq E(G)$. Arbitrarily orient and label each of the edges of G (the orientation need not extend to an orientation of the ribbon graph). The boundary components of the spanning ribbon subgraph $(V(G), A)$ of G meet the edges of G in disjoint arcs (where the spanning ribbon subgraph is naturally embedded in G). On each of these arcs, place an arrow which points in the direction of the orientation of the edge boundary and is labelled by the edge it meets. The resulting marked boundary components of the spanning ribbon subgraph $(V(G), A)$ define an arrow presentation. The ribbon graph corresponding to this arrow presentation is the partial dual $G^{\delta(A)}$ of G.

Example 2.5. Figure 2.7 provides an example of the formation of a partial dual $G^{\delta(A)}$ with $A = \{2, 5\}$ using Definition 2.4. The ribbon graph G is shown in Fig. 2.7a. Figure 2.7b shows the spanning subgraph $(V(G), \{2, 5\})$ with marked boundary components. The arrow presentation for $G^{\delta(A)}$ is shown in Fig. 2.7c. The partial dual $G^{\delta(A)}$ is shown in Fig. 2.7d.

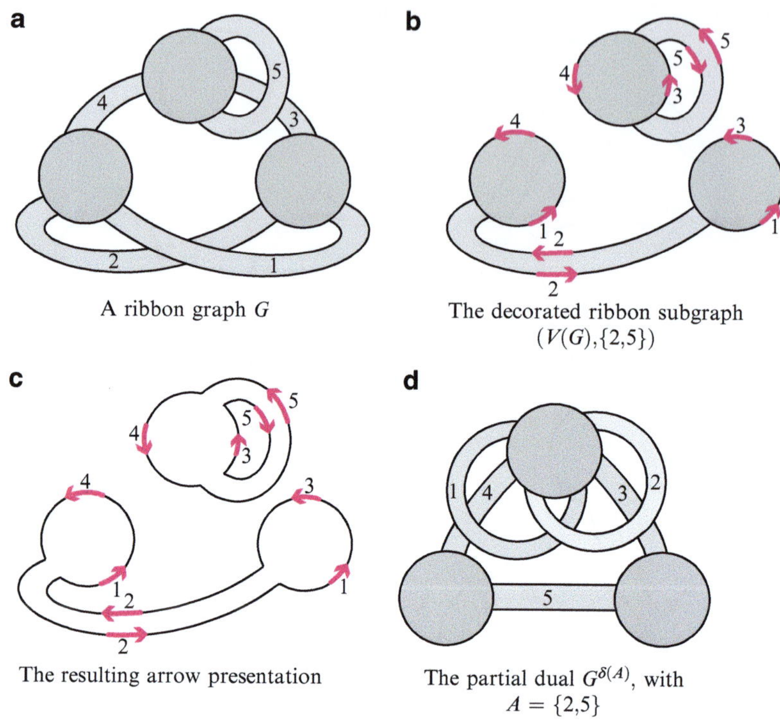

a A ribbon graph G

b The decorated ribbon subgraph $(V(G),\{2,5\})$

c The resulting arrow presentation

d The partial dual $G^{\delta(A)}$, with $A = \{2,5\}$

Fig. 2.7 An example of a partial dual formed through spanning ribbon subgraphs

2.2.2.2 Partial Duals via Ram Graphs

Ram graphs lend themselves to a particularly intuitive construction of partial duals. We would like to think of forming $G^{\delta(A)}$ by taking the usual geometric dual of G, but only at the edges in A. We can realise this construction geometrically by "hiding" the edges that are not in A, that is, by starting with the ram graph which has ribbons for the edges in A and arrows representing the edges not in A, and then taking the geometric dual of the underlying ribbon graph as in the following definition.

Definition 2.6. Given a ribbon graph G and $A \subseteq E(G)$, form the ram graph $G\{A^c\}$. Then form the geometric dual of the ribbon graph $G|_A$ contained in $G\{A^c\}$, but *carrying the arrow markings on the boundaries along in the process*. This results in a new ram graph, which we denote $(G\{A^c\})^{\dagger}$.

Proposition 2.7. *Let G be a ribbon graph and $A \subseteq E(G)$. Then $G^{\delta(A)} = (G\{A^c\})^{\dagger}$.*

Proof. The equivalence between $(G\{A^c\})^{\dagger}$ and $G^{\delta(A)}$ as given in Definition 2.4 (and therefore Definition 2.3) can be seen by observing that $(G\{A^c\})^{\dagger}$ can be obtained from the marked boundary components of the spanning ribbon subgraph $(V(G),A)$ in Definition 2.4 by adding all of the edges corresponding to elements of A. $\qquad\square$

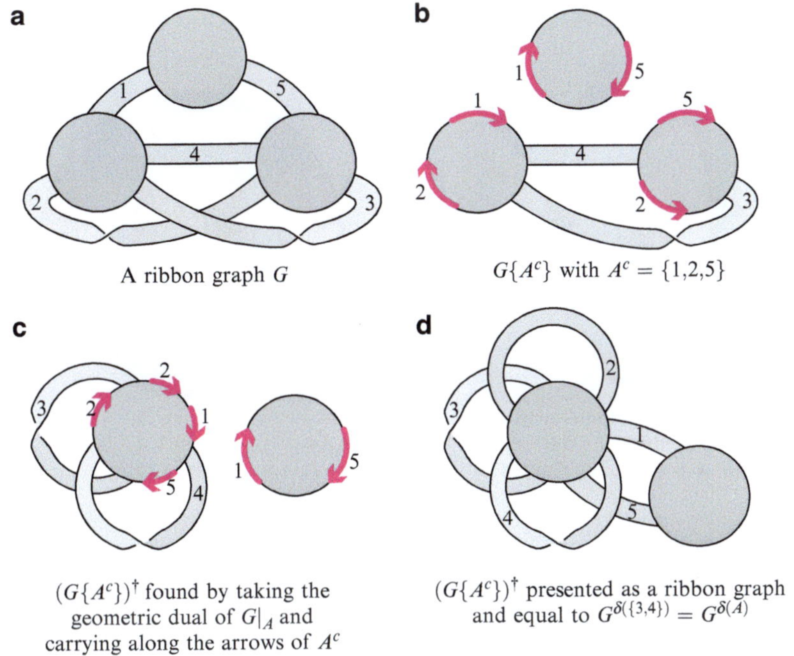

Fig. 2.8 Forming a partial dual using ram graphs

Example 2.8. An example of the formation of a partial dual using Definition 2.6 is given in Figure 2.8.

2.2.2.3 Partial Duals via Ribbon Graphs

The following construction of a partial dual is purely geometric in that it avoids the use of marking arrows.

Definition 2.9 (Bradford, Butler, and Chmutov [8]). Regard the boundary components of the spanning ribbon subgraph $(V(G), A)$ of G as curves on the surface of G. Glue a disc to G along each connected component of this curve and remove the interior of all vertices of G. We denote the result of this operation by $G(A)^\ddagger$.

Proposition 2.10. *Let G be a ribbon graph and $A \subseteq E(G)$. Then $G^{\delta(A)} = G(A)^\ddagger$.*

Proof. The construction in Definition 2.9 can be recovered from that of Definition 2.6 by noting that the edges in A^c just encode the marking arrows. □

Definition 2.9 is particularly pleasing from a conceptual viewpoint as we again see directly that the partial dual $G^{\delta(A)}$ is formed by taking the geometric dual of the ribbon graph G as discussed in Sect. 1.4, but ignoring the edges not in A.

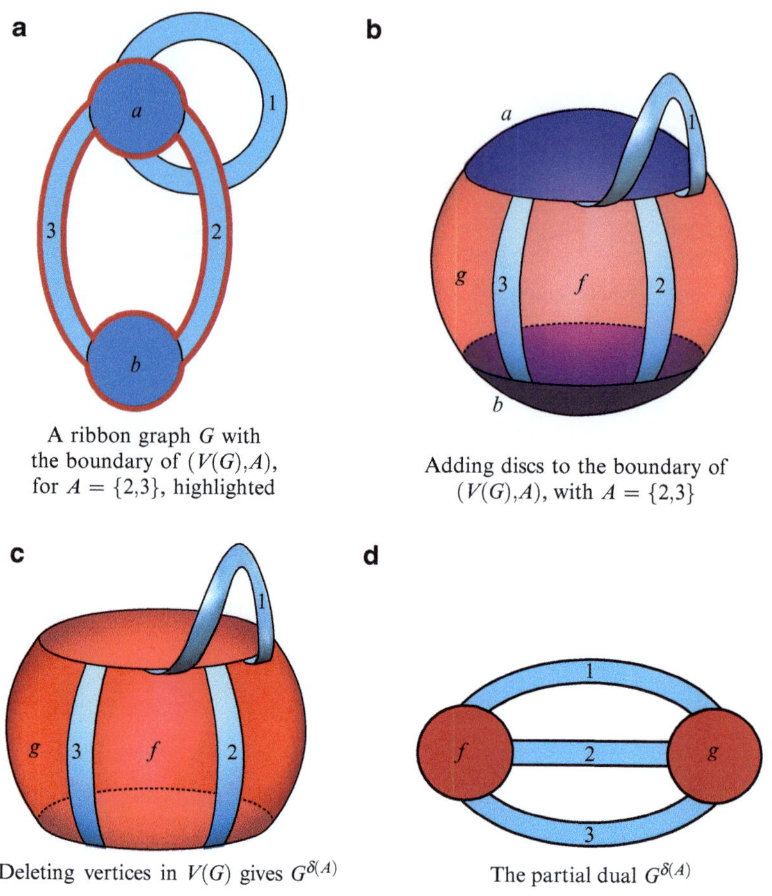

a

A ribbon graph G with
the boundary of $(V(G),A)$,
for $A = \{2,3\}$, highlighted

b

Adding discs to the boundary of
$(V(G),A)$, with $A = \{2,3\}$

c

Deleting vertices in $V(G)$ gives $G^{\delta(A)}$

d

The partial dual $G^{\delta(A)}$

Fig. 2.9 An example of a partial dual formed through ribbon subgraphs

Example 2.11. An example of forming a partial dual $G^{\delta(A)}$ for $A = \{2,3\}$ using Definition 2.9 is given in Fig. 2.9. Figure 2.9b shows the structure obtained by adding discs to G (shown in Fig. 2.9a) that follow the boundary components of $(V(G),A)$, where $A = \{2,3\}$. Figure 2.9c shows the ribbon graph obtained by deleting the vertices of G. The ribbon graph $G^{\delta(A)}$ has been redrawn in Fig. 2.9d.

2.2.2.4 The Partial Dual of an Edge of a Ribbon Graph

Definition 2.3 provides a local construction of partial duals: change part of an arrow presentation as indicated in Fig. 2.4. In contrast, all of the other constructions of partial duals given above are global in that they involve the whole of the embedded graph. Giving local constructions of partial duals in presentations of embedded

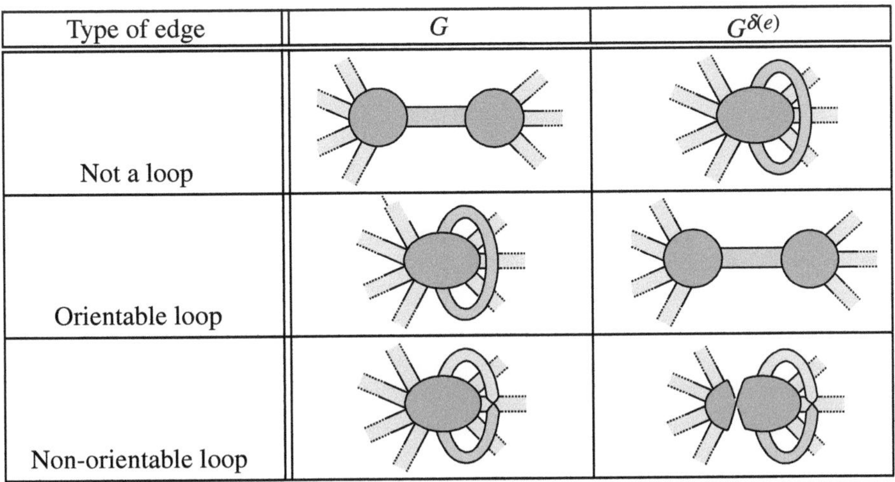

Fig. 2.10 The partial dual of an edge of a ribbon graph

graphs other than arrow presentations can be a little more involved. For example, Fig. 2.10 describes how partial duality at an edge e changes a ribbon graph G. Note that the cases when e is an orientable loop, non-orientable loop, and is not a loop must be treated separately.

2.2.3 Basic Properties of Partial Duality

The following proposition lists a few basic properties of partial duals. Further properties and applications of partial duals will be discussed later in the text.

Proposition 2.12 (Chmutov [16]). *Let G be a ribbon graph and $A, B \subseteq E(G)$. Then the following properties hold:*

1. *$G^{\delta(\emptyset)} = G$.*
2. *$G^{\delta(E(G))} = G^*$, where G^* is the geometric dual of G.*
3. *$(G^{\delta(A)})^{\delta(B)} = G^{\delta(A \triangle B)}$, where $A \triangle B := (A \cup B) \setminus (A \cap B)$ is the symmetric difference of A and B.*
4. *$G^{\delta(\{e,f\})} = (G^{\delta(e)})^{\delta(f)} = (G^{\delta(f)})^{\delta(e)}$. In particular, partial duals can be formed one edge at a time.*

5. *G is orientable if and only if $G^{\delta(A)}$ is orientable.*
6. *Partial duality acts disjointly on components, i.e., $(P \sqcup Q)^{\delta(A)} = (P^{\delta(A \cap E(P))}) \sqcup (Q^{\delta(A \cap E(Q))})$.*
7. *There is a natural 1–1 correspondence between the edges of G and the edges of $G^{\delta(A)}$.*

Proof. Items 1 and 2 follow immediately from Definition 2.6. Item 3 is easily seen by considering the arrow presentation construction of partial duals from Definition 2.1 and observing that $(G^{\delta(e)})^{\delta(e)} = G$. Item 4 follows from Item 3.

For Item 5 we use the construction of Definition 2.9. Suppose G is an orientable ribbon graph and $A \subseteq E(G)$. Adding the discs to the boundary components of $(V(G), A)$ as in Definition 2.9 results in a cell complex X that consists of a (not necessarily connected) closed surface Σ with ribbons attached to it. Let R be this set of ribbons. Σ is orientable since the ribbon subgraph $(V(G), A)$ is. Therefore, if there was a closed curve l in X with a regular neighbourhood $N(l)$ homeomorphic to a Möbius band, it would pass through, in order, the sub-complexes $\Sigma_1 R_1 \Sigma_2 R_2 \cdots R_n \Sigma_1 R_1$ of X, where $\Sigma_i \subseteq \Sigma$ and $R_i \in R$. But as each R_i is an edge in G, and there is a path in Σ_i between R_{i-1} and R_i if and only if there is a path between R_{i-1} and R_i in G, it follows that G contains a curve with a neighbourhood homeomorphic to a Möbius band and is therefore non-orientable. It follows that X is orientable and so $G^{\delta(A)} = X - V(G)$ is. For the converse note that if $G^{\delta(A)}$ is orientable, then, by the above, so is $(G^{\delta(A)})^{\delta(A)} = G$.

Items 6 and 7 are trivial. □

2.3 Twisted Duality

In the preceding sections, we described two operations on embedded graphs: partial Petrality and partial duality. Now we will show how these two operations give rise to a generalisation of duality called twisted duality. The material in this section is based on the results of [30], but here our approach is more geometric than that of [30].

2.3.1 *Sequences of Partial Duals and Petrials*

Let G be an embedded graph and $A \subseteq E(G)$. Also, let $w = w_1 w_2 \cdots w_n$ be a word in the alphabet $\{\delta, \tau\}$. Then we define

$$G^{w(A)} := (\cdots((G^{w_n(A)})^{w_{n-1}(A)}) \cdots)^{w_1(A)}. \tag{2.1}$$

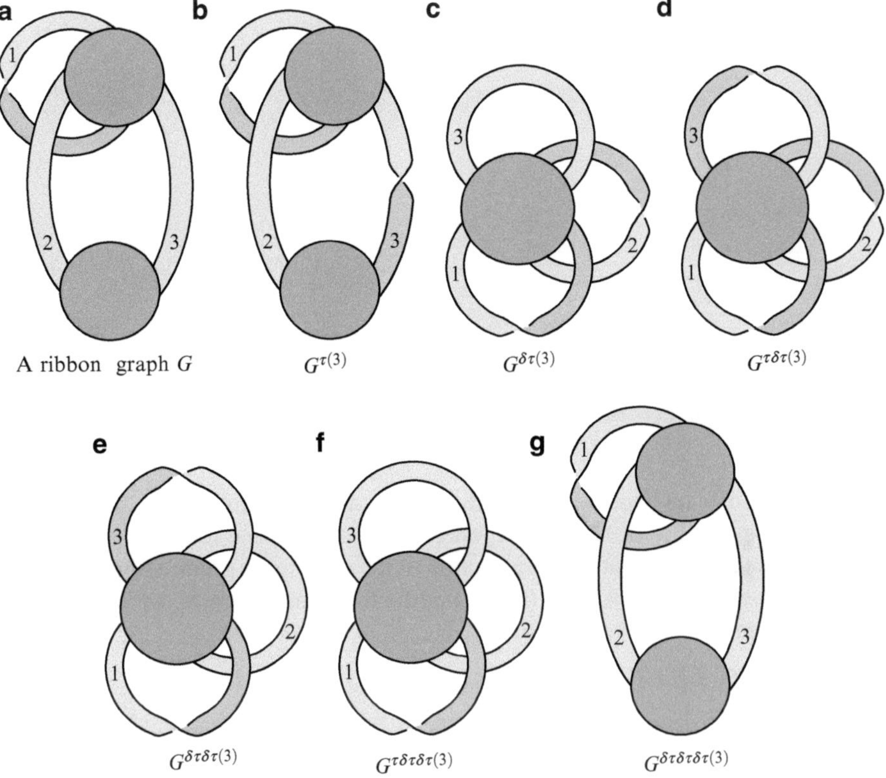

Fig. 2.11 Forming a ribbon graph $G^{\delta\tau\delta\tau\delta\tau(3)}$

In Eq. (2.1), as with function composition, the w_i's in the word w act on A from the right to the left. Also, at each step, we are identifying the edges $A \subseteq E(G)$ with the edges $A \subseteq E\left(\left(\cdots\left(G^{w_n(A)}\right)^{w_{n-1}(A)}\cdots\right)^{w_{n-(i-1)}(A)}\right)$.

Example 2.13. Figure 2.11 illustrates the formation of $G^{w(A)}$ where $w = \delta\tau\delta\tau\delta\tau$ and $A = \{3\}$.

Figure 2.12 shows the actions of δ and τ on an edge of a ribbon graph. Notice that applying some sequences of δ and τ to an edge of G does not change the ribbon graph. This is a consequence of relations among the words in δ and τ:

Proposition 2.14. *Let G be an embedded graph and $A \subseteq E(G)$. Then*

1. $G^{\tau^2(A)} = G$
2. $G^{\delta^2(A)} = G$
3. $G^{(\delta\tau)^3(A)} = G$

Proof. The proposition follows immediately upon considering Fig. 2.12. $\qquad\square$

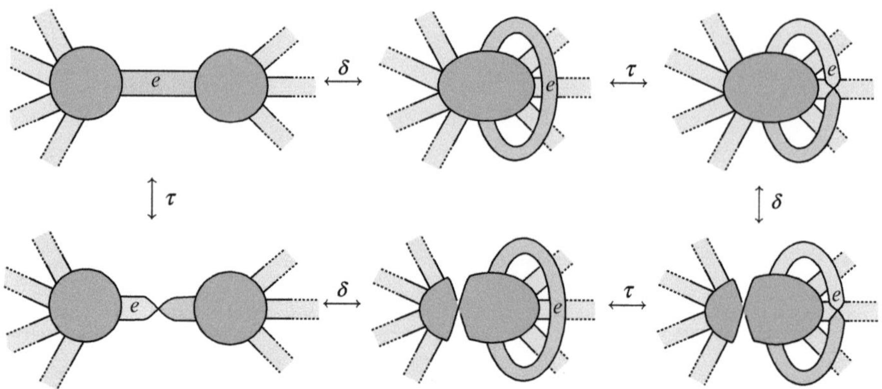

Fig. 2.12 The actions of δ and τ on an edge e of a ribbon graph

The relations on words in δ and τ given in Proposition 2.14 allow us to manipulate a word w in $G^{w(A)}$, but more importantly, these relations allow us to recognise the above operations as an action of the symmetric group S_3 on embedded graphs with specified subsets of edges. For the following theorem, we recall that $\mathscr{G}$ denotes the set of all embedded graphs.

Theorem 2.15. *Let $\mathscr{X} = \{(G,A) \mid G \in \mathscr{G}, A \subseteq E(G)\}$. Then there is a group action of*

$$\mathfrak{G} := \langle \delta, \tau \mid \delta^2, \tau^2, (\delta\tau)^3 \rangle \tag{2.2}$$

on $\mathscr{X}$ given by $g(G,A) := (G^{g(A)}, A)$ for $g \in \mathfrak{G}$. Moreover, the group action is faithful.

Proof. It follows immediately from Eq. (2.1) and Proposition 2.14 that setting $g(G,A) = (G^{g(A)}, A)$ defines a group action of $\mathfrak{G}$ on $\mathscr{X}$. It remains to show that δ and τ satisfy no additional relations. Since $\mathfrak{G}$ is a presentation of the symmetric group S_3 and every proper quotient group of $\mathfrak{G}$ is abelian, it is enough to show that $\tau\delta(G,A) \neq \delta\tau(G,A)$ for some (G,A), a fact that is readily verified. $\square$

When the set A consists of a single edge, or even several specified edges, we sometimes simplify the notation slightly, writing $G^{g(e)}$ for $G^{g(\{e\})}$ and $G^{g(e_1,\dots,e_n)}$ for $G^{g(\{e_1,\dots,e_n\})}$.

2.3.2 Twisted Duals

We have just seen how a word in δ and τ acts on an embedded graph with a fixed subset of edges. In this section we examine what happens when we apply such words to combinations of subsets of edges. This leads to *twisted duality*, which extends all of the forms of duality we have seen so far (i.e., geometric duality, Petrality, partial

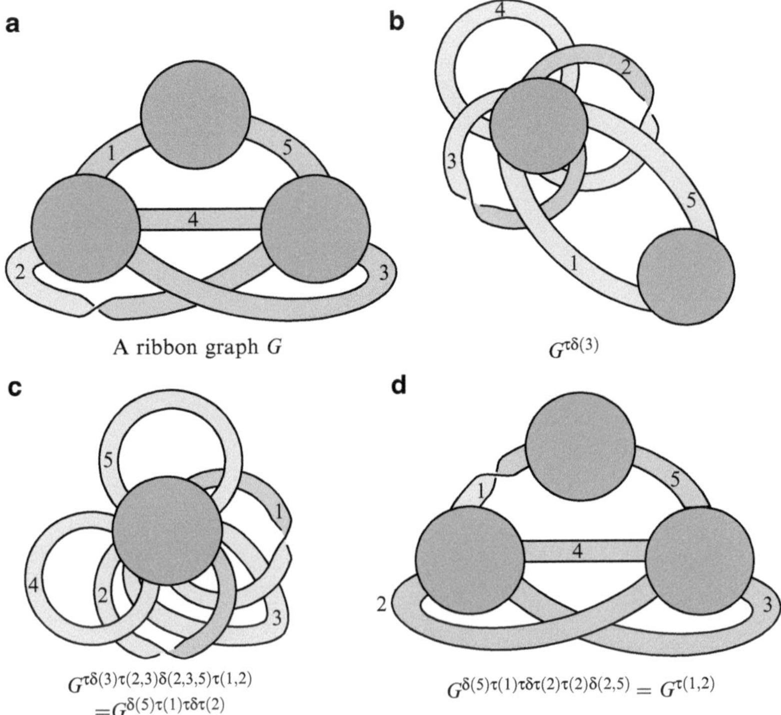

Fig. 2.13 Examples of twisted duals

duality, and partial Petriality). In this section, instead of the formalism of group actions, we focus on operations on embedded graphs, usually presented as ribbon graphs. In Sect. 2.4 we will see how the same twisted duality discussed here can also be obtained through a group action, as it was in [30].

Suppose G is a ribbon graph, $A, B \subseteq E(G)$, and $g, h \in \mathfrak{G} = \langle \delta, \tau \,|\, \delta^2, \tau^2, (\delta\tau)^3 \rangle$. Then we define

$$G^{g(A)h(B)} := \left(G^{g(A)} \right)^{h(B)}.$$

Equipped with this, we are now able to define twisted duality.

Definition 2.16. Embedded graphs G and H are *twisted duals* if there exist $A_1, \ldots, A_n \subseteq E(G)$ and $g_1, \ldots, g_n \in \mathfrak{G}$ such that

$$H = G^{g_1(A_1)g_2(A_2)\cdots g_n(A_n)}.$$

Consistent with our previous terminology we can regard the act of taking a twisted dual as an operation which we call twisted duality.

Figure 2.13 shows the formation of three twisted duals of a ribbon graph G. The set of twisted duals of the plane 2-cycle is shown in Fig. 2.14.

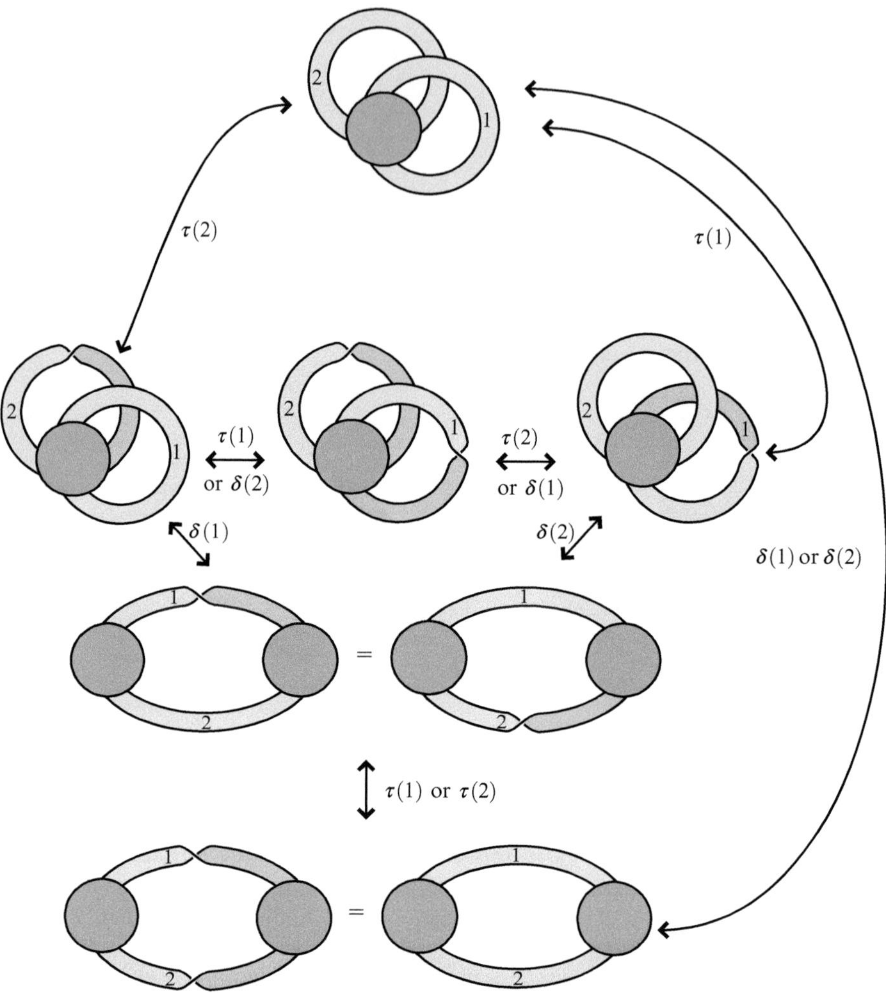

Fig. 2.14 All the twisted duals of the 2-cycle

As can be seen in Figs. 2.13 and 2.14, twisted duals of the same embedded graph can have very different graph theoretical and topological properties from one another. Also, although an embedded graph can have at most $6^{|E(G)|}$ twisted duals many of these twisted duals may be equivalent.

All of the identities in Proposition 2.14 for a single edge also hold for twisted duals, i.e., the group relations hold when applied to a set of edges A. However, when the exponent in a twisted dual involves more than one set, additional identities also hold.

Proposition 2.17. *Let G be an embedded graph, $A, B \subseteq E(G)$ and $g, h \in \mathfrak{G}$. Then the following hold:*

1. *If $A \cap B = \emptyset$, then $G^{g(A)h(B)} = G^{h(B)g(A)}$.*
2. *$G^{g(A)} = (G^{g(e)})^{g(A \setminus \{e\})}$, when $e \in A$, and so twisted duals can be formed one edge at a time.*
3. *Twisted duality acts disjointly on the components of G.*
4. *There is a natural 1–1 correspondence between the edges of G and the edges of each of its twisted duals.*

Proof. These identities hold since they hold for partial Petrials (by Proposition 2.2) and partial duals (by Proposition 2.12). $\qquad\square$

Propositions 2.14 and 2.17 allow us to express twisted duals in a generic form, as follows.

Proposition 2.18. *If, for all i belonging to some indexing set, we have $h_i \in \mathfrak{G}$ and $B_i \subseteq E(G)$, then any expression of the form $G^{\prod h_i(B_i)}$, where the product is over the i's, can be written as*

$$G^{1(A_1)\delta(A_2)\tau(A_3)\tau\delta(A_4)\delta\tau(A_5)\tau\delta\tau(A_6)}, \tag{2.3}$$

where the A_i partition $E(G)$. Moreover, this expression is unique.

Proof. The proposition follows by repeated applications of Propositions 2.14 and 2.17. $\qquad\square$

Figure 2.13 illustrates how the exponents can be written in the generic form from Proposition 2.18.

Note that, in the notation of Proposition 2.18, the terms in the product $\prod h_i(B_i)$ do not necessarily commute, while the terms in the product $\prod_{i=1}^{6} g_i(A_i)$ do commute with each other since the A_i's are disjoint.

Notation 2.19. We use G^{Γ} to denote a generic expression of the form $G^{g_1(A_1)g_2(A_2)\cdots g_n(A_n)}$, where each $g_i \in \mathfrak{G}$ and $A_i \subseteq E(G)$. We may assume without loss of generality that G^{Γ} is of the form shown in Eq. (2.3), and we can omit factors of the form $1(A_1)$ or $g_i(\emptyset)$.

It is useful to have a taxonomy, given in Table 2.1, for "partial" analogues of other derived graphs such as the Wilson dual (or opposite graph), trialities, and direct derivatives from, for example, [100], especially since these, like partial duals and partial Petrials, arise from applying group elements of $\mathfrak{G}$ to embedded graphs. We also note that when $A = E(G)$, the result is the action of $\mathfrak{G}$ on an embedded graph (usually a regular map) studied by Wilson in [100].

2.4   The Ribbon Group and its Action

In Definition 2.16, twisted duals were formed by applying one of the six elements of $\mathfrak{G}$ to specified subsets of the edges. However, twisted duals can equivalently be defined by a group action of $\mathfrak{G}^n$ on embedded graphs with n edges. The significance

Table 2.1 Taxonomy of classes of twisted duals

Generator(s)	Order of subgroup of $\mathfrak{G}$ generated	Applied to all edges	Applied to a subset of edges
δ	2	Geometric dual	Partial dual
τ	2	Petrie dual or Petrial	Partial Petrial
$\tau\delta\tau$	2	Wilson dual or Wilsonial (or the opposite)	Partial Wilsonial
$\delta\tau$	3	Triality	Partial triality
δ and τ	6	A direct derivative	Twisted dual

of this approach is that the twisted duals of an embedded graph G, with an arbitrary order on its edges, form the orbit of G under this group action, while geometric duality, Petrials, partial duality, and partial Petrials, as well as the trialities, partial trialities, Wilsonials, and partial Wilsonials described above, arise as the orbits of various subgroups of $\mathfrak{G}^n$.

2.4.1 Defining the Group Action

By Theorem 2.15, the group $\mathfrak{G} = \langle \delta, \tau \,|\, \delta^2, \tau^2, (\delta\tau)^3 \rangle$ acts on the set of ribbon graphs with a distinguished edge by $g(G, \{e\}) = (G^{g(e)}, \{e\})$. We now extend this construction to an action of the direct product $\mathfrak{G}^n$ of n copies of $\mathfrak{G}$ on the set of embedded graphs with n linearly ordered edges by making each factor in the direct product act on a specific edge of an embedded graph.

Definition 2.20. We call $\mathfrak{G}^n$ the *ribbon group for embedded graphs with n edges*.

To define the action of the ribbon group, we work with embedded graphs with arbitrarily ordered edges. Recall that $\mathscr{G}_n$ denotes the set of embedded graphs with exactly n edges. Let

$$\mathscr{G}_{\mathrm{or}(n)} = \{(G, \prec) \,|\, G \in \mathscr{G}_n \text{ and } \prec \text{ is a linear ordering of } E(G)\}$$

denote the set of embedded graphs with exactly n linearly ordered edges.

Definition 2.21. We define the *ribbon group action* of the ribbon group on $\mathscr{G}_{\mathrm{or}(n)}$ as follows. Given $(g_1, g_2, \ldots, g_n) \in \mathfrak{G}_n$ and $(G, e_1 \prec e_2 \prec \cdots \prec e_n) \in \mathscr{G}_n$, then

$$(g_1, g_2, \ldots, g_n)(G, e_1 \prec e_2 \prec \cdots \prec e_n) = (G^{g_1(e_1)g_2(e_2)\cdots g_n(e_n)}, e_1 \prec e_2 \prec \cdots \prec e_n).$$

$$(2.4)$$

The following theorem states that the ribbon group action does indeed define a group action on $\mathscr{G}_{\mathrm{or}(n)}$.

Theorem 2.22. *The action of $\mathfrak{G}^n$ on $\mathcal{G}_{\mathrm{or}(n)}$ defined in Eq. (2.4) is a group action.*

Proof. The result follows easily from Theorem 2.15.   □

We record here, without proof, the following result from [30] which provides some properties of the ribbon group action.

Proposition 2.23. *The ribbon group action of $\mathfrak{G}^n$ on $\mathcal{G}_{\mathrm{or}(n)}$ is*

1. *Faithful*
2. *Has no fixed points*
3. *Transitive if and only if $n > 1$*
4. *Not free*

2.4.2   Recovering Dualities from Actions of Subgroups of the Ribbon Group

Here we see how twisted duality and its various specialisations arise as orbits under the action of the ribbon group and its subgroups.

Suppose that a group $\mathfrak{H}$ acts on a set X. Recall that the *orbit* of an element $x \in X$ is the set

$$\mathrm{Orb}_{\mathfrak{H}}(x) := \{hx \,|\, h \in \mathfrak{H}\}.$$

It turns out that twisted duals arise from the orbits of embedded graphs under the ribbon group action.

The ribbon group $\mathfrak{G}^n$ acts on the set $\mathcal{G}_{\mathrm{or}(n)}$ of embedded graphs with n ordered edges. To make contact with twisted duality and its specialisations, which do not require an edge order, we make the following definition.

Definition 2.24. Let G be an embedded graph and $\mathfrak{H}$ be a subgroup of the ribbon group $\mathfrak{G}^n$. Then we define $\mathrm{Orb}_{\mathfrak{H}}(G)$ to be the set of embedded graphs (without edge orders) in $\mathrm{Orb}_{\mathfrak{H}}(G, \prec)$ where $\prec$ is any edge order of G.

In Definition 2.24, the action of $\mathfrak{H}$ on $\mathcal{G}_{\mathrm{or}(n)}$ is the one arising from the ribbon group action.

We can now recover all of the dualities discussed so far via orbits of embedded graphs under the action of the ribbon group and its subgroups.

Theorem 2.25. *Let G and H be embedded graphs.*

1. *G and H are geometric duals if and only if $H \in \mathrm{Orb}_{\mathfrak{H}}(G)$, where $\mathfrak{H}$ is the subgroup of $\mathfrak{G}^n$ generated by $(\delta, \delta, \ldots, \delta)$.*
2. *G and H are partial duals if and only if $H \in \mathrm{Orb}_{\mathfrak{H}}(G)$, where $\mathfrak{H}$ is the subgroup of $\mathfrak{G}^n$ generated by $\{(\delta, 1, \ldots, 1), (1, \delta, 1, \ldots, 1), \ldots, (1, 1, \ldots, \delta)\}$.*
3. *G and H are Petrials if and only if $H \in \mathrm{Orb}_{\mathfrak{H}}(G)$, where $\mathfrak{H}$ is the subgroup of $\mathfrak{G}^n$ generated by $(\tau, \tau, \ldots, \tau)$.*
4. *G and H are partial Petrials if and only if $H \in \mathrm{Orb}_{\mathfrak{H}}(G)$, where $\mathfrak{H}$ is the subgroup of $\mathfrak{G}^n$ generated by $\{(\tau, 1, \ldots, 1), (1, \tau, 1, \ldots, 1), \ldots, (1, 1, \ldots, \tau)\}$.*

5. *G and H are Wilsonials if and only if $H \in \mathrm{Orb}_{\mathfrak{H}}(G)$, where $\mathfrak{H}$ is the subgroup of $\mathfrak{G}^n$ generated by $(\tau\delta\tau, \tau\delta\tau, \ldots, \tau\delta\tau)$.*
6. *G and H are partial Wilsonials if and only if $H \in \mathrm{Orb}_{\mathfrak{H}}(G)$, where $\mathfrak{H}$ is the subgroup of $\mathfrak{G}^n$ generated by $\{(\tau\delta\tau, 1, \ldots, 1), (1, \tau\delta\tau, 1, \ldots, 1), \ldots, (1, 1, \ldots, \tau\delta\tau)\}$.*
7. *G and H are trialities if and only if $H \in \mathrm{Orb}_{\mathfrak{H}}(G)$, where $\mathfrak{H}$ is the subgroup of $\mathfrak{G}^n$ generated by $(\delta\tau, \delta\tau, \ldots, \delta\tau)$.*
8. *G and H are partial trialities if and only if $H \in \mathrm{Orb}_{\mathfrak{H}}(G)$, where $\mathfrak{H}$ is the subgroup of $\mathfrak{G}^n$ generated by $\{(\delta\tau, 1, \ldots, 1), (1, \delta\tau, 1, \ldots, 1), \ldots, (1, 1, \ldots, \delta\tau)\}$.*
9. *G and H are direct derivatives if and only if $H \in \mathrm{Orb}_{\mathfrak{H}}(G)$, where $\mathfrak{H}$ is the subgroup of $\mathfrak{G}^n$ generated by $\{(\delta, \delta, \ldots, \delta), (\tau, \tau, \ldots, \tau)\}$.*
10. *G and H are twisted duals if and only if $H \in \mathrm{Orb}_{\mathfrak{G}^n}(G)$.*

Proof. The theorem follows by the construction of the ribbon group action.   □

Notation 2.26. Because of the frequency of its use, we will write simply $\mathrm{Orb}(G)$ for $\mathrm{Orb}_{\mathfrak{G}^n}(G)$, the set of all twisted duals of an embedded graph G. In addition, if $g \in \mathfrak{G}$, we define

$$\mathrm{Orb}_{(g)}(G) := \mathrm{Orb}_{\mathfrak{H}}(G),$$

where $\mathfrak{H}$ is the subgroup of $\mathfrak{G}^n$ generated by $\{(g, 1, \ldots, 1), (1, g, 1, \ldots, 1), \ldots, (1, 1, \ldots, g)\}$.

Chapter 3
Twisted Duality, Cycle Family Graphs, and Embedded Graph Equivalence

In Sect. 1.5.3 we saw that geometric duals completely characterise the set of embedded graphs with a given embedded medial graph. This result was obtained by using Tait graphs, which describe all of the embedded graphs with a given embedded medial graph. In fact, Sect. 1.5.3 provided fundamental connections among the three concepts of a medial graph, a geometric dual, and a Tait graph. In this chapter, we extend these connections to twisted duality. In particular, we describe how twisted duality and partial duality solve isomorphism problems about medial graphs. In doing so we obtain a hierarchical understanding of how types of duality correspond to types of embedded graph equivalence.

We begin by recalling, from Item 5 of Theorem 1.15, that if G is a plane graph, then G and G^* are the only plane graphs with the same medial graph as G, i.e.,

$$\{G, G^*\} = \{H \mid H_m = G_m\}. \tag{3.1}$$

Some observations emerge from considering Eq. (3.1).

Observation 1: Medial graphs, Tait graphs, and geometric duals generate one another. In particular:

1. Given a 4-regular plane graph F, the Tait graphs of F are precisely the embedded graphs with medial graph F.
2. If G_m is a plane medial graph, then the Tait graphs of G_m are exactly the embedded graphs with medial graphs equivalent to G_m, and the two Tait graphs are geometric duals. This means G and G^* are exactly the Tait graphs of G_m.
3. Plane graphs G and G^* are precisely the plane graphs with the same plane medial graph.

This leads to the first question.

Question 3.1. What is the twisted duality analogue of the connections among geometric duality, Tait graphs, and medial graphs? In particular, how does Theorem 1.15 extend to twisted duality?

J.A. Ellis-Monaghan and I. Moffatt, *Graphs on Surfaces: Dualities, Polynomials, and Knots*, SpringerBriefs in Mathematics, DOI 10.1007/978-1-4614-6971-1_3,

The second observation concerns the nature of the objects and relations in Eq. (3.1).

Observation 2: There is a correspondence between geometric duality and equivalence of embedded graphs, in that Eq. (3.1) is specific to *geometric* duality and *equality* of embedded graphs. In particular:

1. If we start with equality of embedded graphs, then the right-hand side of Eq. (3.1) generates geometric duals. That is, G^* results from finding all the embedded graphs with medial graphs equivalent as embedded graphs to the medial graph of G.
2. If we start with geometric duality, then a graph and its geometric dual have medial graphs which are equivalent as embedded graphs. That is, if we know that two graphs are geometric duals, then we immediately know that their medial graphs are equivalent as embedded graphs.

This leads to the second question.

Question 3.2. What types of dualities are generated when we change the type of graph equivalence among medial graphs, and what forms of graph equivalence are generated when we change our concept of duality? More specifically, how is a hierarchy of graph equivalences captured by a hierarchy of dualities and vice versa?

To answer Question 3.1, we construct cycle family graphs as full generalisations of Tait graphs to embedded graphs, showing that they characterise all twisted duals with a given (abstract) medial graph. For Question 3.2, we give a hierarchy of graph structures, ranging from abstract graphs, through cyclically ordered graphs, to embedded graphs. We then use these structures to define a hierarchy of equivalences of embedded graphs. We show that when we consider medial graphs up to equivalence under these varying levels of specificity, then there is a corresponding hierarchy of duality, ranging from twisted duality, through partial duality, to geometric duality.

3.1   Characterising Orb(*G*)

This section establishes the necessary structures and theory to answer Question 3.1. We determine the complete twisted duality analogue to the relations among medial graphs, Tait graphs, and geometric duality. In Theorem 3.8, we see that cycle family graphs, a full generalisation of Tait graphs, precisely characterise the twisted duals of an embedded graph. We also show that the ribbon group action gives a complete characterisation of twisted duality in that if G is any cellularly embedded graph with medial graph G_m, then the orbit of G under the group action is precisely the set of all embedded graphs with medial graphs equivalent (as abstract graphs) to G_m. Thus, the results of this section fully generalise the properties of Theorem 1.15 for plane graphs to graphs embedded in any surface.

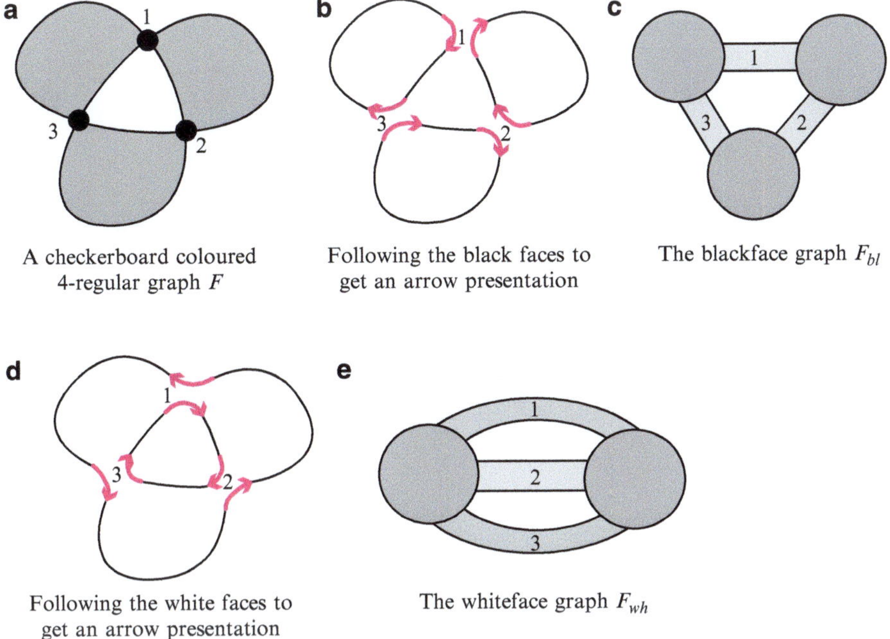

a — A checkerboard coloured 4-regular graph F

b — Following the black faces to get an arrow presentation

c — The blackface graph F_{bl}

d — Following the white faces to get an arrow presentation

e — The whiteface graph F_{wh}

Fig. 3.1 Forming Tait graphs using arrow presentations

3.1.1 Extending Tait Graphs to Cycle Family Graphs

We begin by extending Tait graphs to cycle family graphs. To motivate this, we begin by considering the construction of Tait graphs in terms of arrow presentations. Let F be a checkerboard coloured plane graph. The Tait graph F_{bl} is found (as in Sect. 1.5.1) by placing a vertex in each black face and adding an edge between vertices wherever the black faces meet. We can describe this construction in terms of arrow presentations as follows: form the circles of the arrow presentation by following the boundary of each black face. Place a pair of c-labelled arrows on the two arcs of the circles that are nearest to the crossing c. Each arrow is directed clockwise about the black face on whose boundary it lies. The Tait graph F_{wh} is constructed in the same way, but using the white faces instead of the black faces. These constructions are illustrated in Fig. 3.1, which should be compared to Fig. 1.14. (Note that in Fig. 3.1b, c, for clarity, we are using a single integer to denote the labels of two arrows.)

Notice that when using arrow presentations, Tait graphs can be obtained by a local move that replaces each vertex v, as in Fig. 3.2a, with the smoothing shown in Fig. 3.2b for the blackface graph, and in Fig. 3.2c for the whiteface graph. Thus, we obtain a Tait graph by taking a checkerboard colouring and smoothing each vertex as in Fig. 3.2.

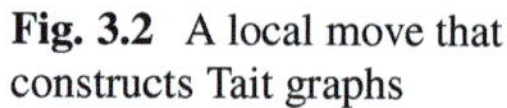

Fig. 3.2 A local move that constructs Tait graphs

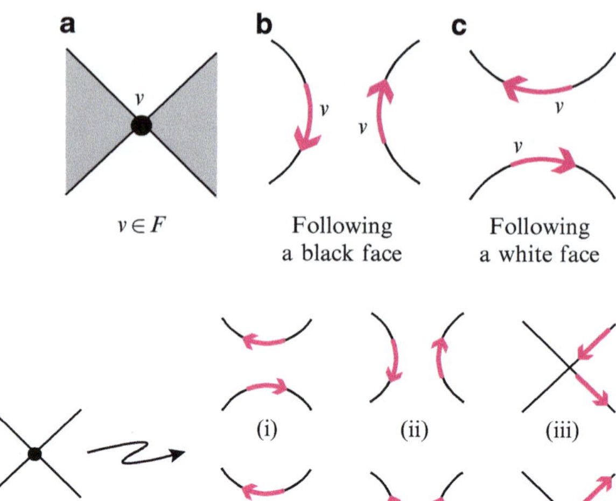

Fig. 3.3 The six arrow marked vertex states used to construct cycle family graphs

We can generalise Tait graphs by removing the restriction that the smoothings of Fig. 3.2 follow a checkerboard colouring. We observe that: (1) there are three, rather than two, possible ways to pair the half-edges incident to a vertex (the two smoothings and a crossing); (2) the labelled arrows can agree or disagree with a local orientation of a vertex; and (3) we need not choose the same type of configuration at every vertex. Thus, we can extend the idea of Tait graphs by removing these restrictions as in the following definition.

Definition 3.3. Let F be a 4-regular cellularly embedded graph. A *cycle family graph* of F is an embedded graph obtained as the arrow presentation given by replacing each vertex with one of the six *arrow marked vertex states* shown in Fig. 3.3. The states labelled (i) and (ii) are *consistent smoothings*, while those labelled (iv) and (v) are *inconsistent smoothings*. Similarly, the state labelled (iii) is a *consistent crossing* and that labelled (vi) is an *inconsistent crossing*. We let $\mathscr{C}(F)$ denote the set of cycle family graphs of F.

Figure 3.4 illustrates the construction of one of the cycle family graphs of a graph in the real projective plane. Figure 3.5 gives an example of the set of all cycle family graphs of a 4-regular embedded graph.

The following special subset of cycle family graphs will be important later, and the reason for using δ in the notation will become apparent.

Definition 3.4. Let F be a 4-regular cellularly embedded graph. A *duality state* of F is a state $\vec{s}$ given by replacing each vertex with one of the two consistent smoothings labelled (i) and (ii) in Fig. 3.3. A *smoothing graph* of F is an embedded graph obtained as the arrow presentation resulting from a duality state. We let $\mathscr{C}_{(\delta)}(F)$ denote the set of smoothing graphs of F.

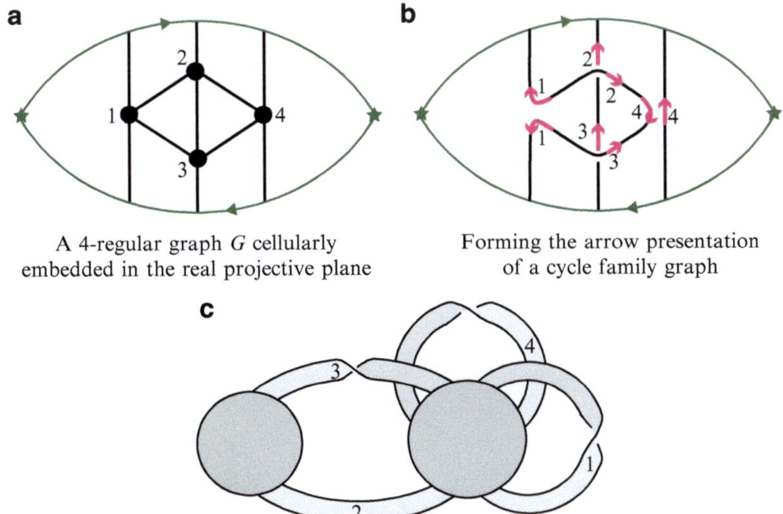

A 4-regular graph G cellularly embedded in the real projective plane

Forming the arrow presentation of a cycle family graph

The cycle family graph as a ribbon graph

Fig. 3.4 Forming a cycle family graph

3.1.2 *Twisted Duality and Cycle Family Graphs*

Having equipped ourselves with cycle family graphs, which generalise the concept of Tait graphs, we can now systematically prove the analogues of Theorem 1.15 (which relates Tait graphs and medial graphs). We begin with the key property of Tait graphs, from Theorem 1.15 Item 1, that if F is a 4-regular embedded graph, then

$$\{F_{bl}, F_{wh}\} = \{G \,|\, G_m = F\}. \tag{3.2}$$

Theorem 3.5. *Let F be a 4-regular abstract graph and let $\tilde{F}$ be any embedding of F. Then*

$$\mathscr{C}(\tilde{F}) = \{G \,|\, G_m \cong F\},$$

i.e.,

$$G_m \cong F \iff G \in \mathscr{C}(\tilde{F}),$$

for some embedding $\tilde{F}$ of F.

Proof. We first show that if $G \in \mathscr{C}(\tilde{F})$, then $G_m \cong F$. The underlying abstract graph of G_m, where G is a cycle family graph of $\tilde{F}$, can be constructed as follows.

- G is obtained by, for each vertex v of $\tilde{F}$, choosing an arrow marked state from Fig. 3.3.

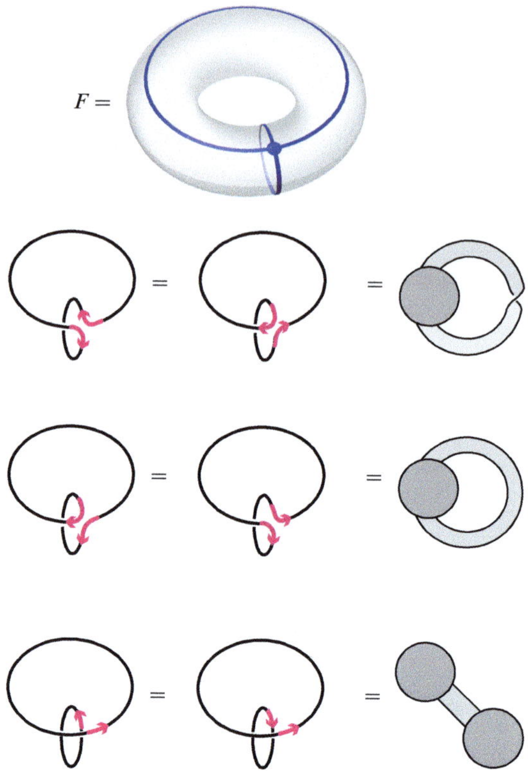

Fig. 3.5 Forming a full set of cycle family graphs, $\mathscr{C}(F)$

- At each pair of v-labelled arrows in the resulting arrow presentation, replace the arrows with a 4-valent vertex as follows: add a vertex; connect this vertex to the arrow presentation by adding an arc between the vertex and the tip and tail of each v-labelled arrow.
- Delete all of the arrows and the arcs on which they lie from the resulting diagram.

This construction is illustrated in Fig. 3.6.

This results in a 4-regular abstract graph that is clearly the underlying abstract graph of the medial graph G_m. For the arrow marked state shown in Fig. 3.6b, we immediately see that the vertex v in $\tilde{F}$ and the corresponding vertex in G_m have the same adjacency information (we are only interested in abstract graphs, so we do not care about the cyclic order of half-edges at the vertices). Moreover, doing the same calculation as in Fig. 3.6 with the other five arrow marked vertex states from Fig. 3.3 that are used in the construction of cycle family graphs also results in the vertex v in $\tilde{F}$ and the corresponding vertex in G_m having the same adjacency information. Thus G_m and $\tilde{F}$ have the same underlying abstract graphs, giving that $G_m \cong F$, as required.

Conversely, suppose that G is an embedded graph such that $G_m \cong F$. Then $G_m \cong \tilde{F}$. Consider an edge e and think of G as both a ribbon graph and an arrow

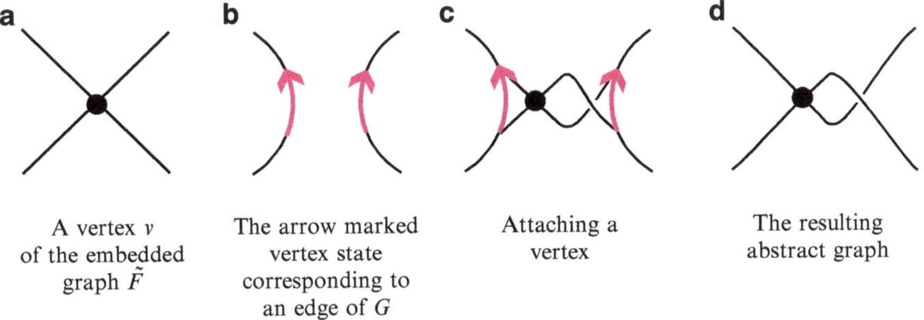

A vertex v of the embedded graph $\tilde{F}$

The arrow marked vertex state corresponding to an edge of G

Attaching a vertex

The resulting abstract graph

Fig. 3.6 Forming the underlying abstract graph of G_m

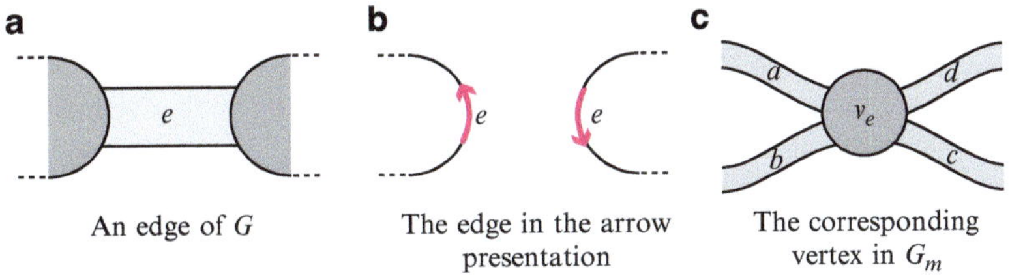

An edge of G

The edge in the arrow presentation

The corresponding vertex in G_m

Fig. 3.7 An edge of a ribbon graph and the corresponding vertex of its medial graph

presentation as in Fig. 3.7a, b, respectively. (We may assume the edge e is as drawn in the figure by pushing any twists to vertices as in Fig. 1.4 if necessary.) Consider also the vertex v_e of G_m that corresponds to e and suppose the edges incident to v_e are in the cyclic order $(a\,b\,c\,d)$ as shown in Fig. 3.7c.

Using that $\tilde{F} \cong G_m$, let $\tilde{v}$ be the vertex of $\tilde{F}$ identified with v_e under the equivalence, and let $\tilde{a}, \tilde{b}, \tilde{c}, \tilde{d}$ be the edges corresponding to a, b, c, d, respectively. Then $\tilde{F}$ and G_m can only differ at $\tilde{v}$ and v_e in the following ways: (1) the number of half-twists in the edges, and (2) the cyclic order of the incident edges.

We need to show that G arises as a cycle family graph of $\tilde{F}$. That is, we need to show that regardless of the number of half-twists and the cyclic order at $\tilde{v}$, the state shown in Fig. 3.7b always appears as an arrow marked state of a cycle family graph of F. To do this we first observe that adding and removing half-twists to the edges $\tilde{a}, \tilde{b}, \tilde{c}, \tilde{d}$ does not affect the equivalence class of the cycle family graph given by a particular set of arrow marked vertex states. Secondly, the cyclic order of the half-edges incident to $\tilde{v}$ is either $(\tilde{a}\tilde{b}\tilde{c}\tilde{d})$, $(\tilde{a}\tilde{b}\tilde{d}\tilde{c})$, $(\tilde{a}\tilde{c}\tilde{b}\tilde{d})$, $(\tilde{a}\tilde{c}\tilde{d}\tilde{b})$, $(\tilde{a}\tilde{d}\tilde{b}\tilde{c})$, or $(\tilde{a}\tilde{d}\tilde{c}\tilde{b})$. However, adding half-twists to edges does not change the cycle family graphs arising from an arrow marked state, and the cyclic order at $\tilde{v}$ can be reversed by flipping the vertex, as in Fig. 1.4, and removing the half-twist this creates from each edge. Thus it follows that $\tilde{F}$ and G_m only differ at $\tilde{v}$ and v_e by the number of half-twists of incident edges and the cyclic order, but we can assume that the cyclic order at $\tilde{v}$ is one of $(\tilde{a}\tilde{b}\tilde{c}\tilde{d})$, $(\tilde{a}\tilde{b}\tilde{d}\tilde{c})$, or $(\tilde{a}\tilde{c}\tilde{b}\tilde{d})$. Figure 3.8 then shows the appropriate choice of arrow marked vertex state to choose at $\tilde{v}$ to recover G as the cycle family graph, completing the proof. $\square$

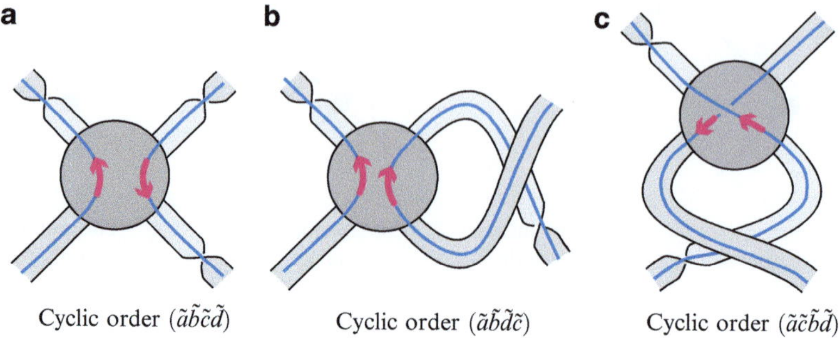

Fig. 3.8 Recovering G as a cycle family graph of $\tilde{F} \cong G_m$

Given a 4-regular abstract graph F, Theorem 3.5 provides a way to construct *every* embedded graph whose medial graph is an embedding of F: given the abstract graph F, choose any embedding $\tilde{F}$ of F and construct its set of cycle family graphs $\mathscr{C}(\tilde{F})$. Note that although not all embeddings of F arise in this way, all the checkerboard colourable embeddings of F (i.e., those that are medial graphs) do.

Example 3.6. If F is the abstract graph with one vertex and two edges, then the embedded graph in Fig. 3.5 is an embedding of F, and the three ribbon graphs shown in Fig. 3.5 give the complete set of ribbon graphs each with a medial graph that is an embedding of F.

We get the analogue of Item 2 of Theorem 1.15 (which states that $(F_{bl})_m = (F_{wh})_m = F$) as an immediate corollary.

Corollary 3.7. *Let F be a 4-regular embedded graph and $\vec{s}$ be an arrow marked graph state of F. Then $(F_{\vec{s}})_m$ and F are equivalent as abstract graphs, i.e., $(F_{\vec{s}})_m \cong F$.*

Proof. $F_{\vec{s}} \in \mathscr{C}(F)$, but by Theorem 3.5, $\mathscr{C}(F) = \{G \,|\, G_m \cong F\}$, so $(F_{\vec{s}})_m \cong F$. □

We can now give the analogue of Item 3 of Theorem 2.22, which states that $\{(G_m)_{bl}, (G_m)_{wh}\} = \{G, G^*\}$. Theorem 3.8 gives the corresponding result for twisted duality, showing that cycle family graphs exactly characterise twisted duals and also gives an alternative construction for them.

Theorem 3.8. *Let G be an embedded graph. Then the cycle family graphs of its medial graph G_m are exactly its twisted duals, i.e.,*

$$\mathscr{C}(G_m) = \mathrm{Orb}(G).$$

Proof. Let $H \in \mathrm{Orb}(G)$. Then the arrow presentation at each edge of H is of one of the types shown in Table 3.1. But these are exactly the arrow presentations of the cycle family graphs of G_m that arise by replacing each vertex of G_m with one of the arrow marked states shown in Fig. 3.3. Thus $H \in \mathscr{C}(G_m)$.

Table 3.1 Arrow presentations of twisted duals at an edge e

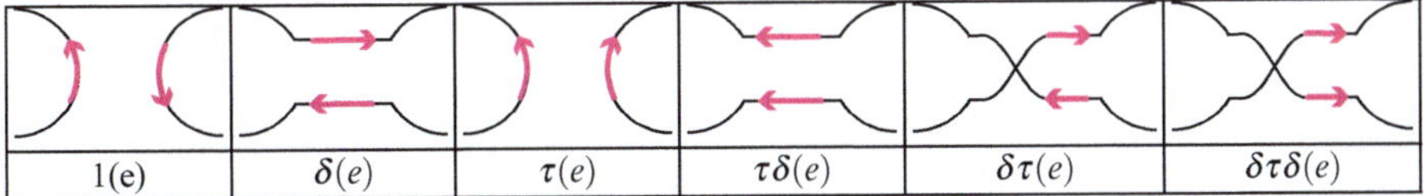

1(e)	$\delta(e)$	$\tau(e)$	$\tau\delta(e)$	$\delta\tau(e)$	$\delta\tau\delta(e)$

Conversely, if $H \in \mathscr{C}(G_m)$, then it is obtained by replacing each vertex of G_m with one of the arrow marked states shown in Fig. 3.3. But these are exactly the arrow presentations of the twisted duals of G as shown in Table 3.1, so $H \in \mathrm{Orb}(G_m)$, completing the proof. $\square$

Theorem 3.8 states that we can obtain all of the cycle family graphs of a medial graph as a set of twisted duals. For example, if G is the non-orientable ribbon graph consisting of one vertex and one edge, then its medial graph G_m consists of a single vertex and two alternating non-orientable loops. It is easily verified that the cycle family graphs $\mathscr{C}(G_m)$ are exactly the twisted duals of G.

Combining Theorems 3.5 and 3.8 yields the analogue of Theorem 1.15, Item 4, which states that $F_{bl} = (F_{wh})^{*}$.

Corollary 3.9. *If F is a 4-regular embedded graph and $F_{\vec{s}}$ and $F_{\vec{s}'}$ are two cycle family graphs of any embedding of F, then $F_{\vec{s}}$ and $F_{\vec{s}'}$ are twisted duals, that is,*

$$F_{\vec{s}} = (F_{\vec{s}'})^{\Gamma}$$

for some Γ.

Proof. By Theorem 3.5,

$$F_{\vec{s}}, F_{\vec{s}'} \in \mathscr{C}(F) = \{G \mid G_m \cong F\}.$$

Thus, $(F_{\vec{s}})_m \cong F \cong (F_{\vec{s}'})_m$. Then, also by Theorem 3.5, $\mathscr{C}(F_{\vec{s}}) = \mathscr{C}(F_{\vec{s}'})$. Finally, using Theorem 3.8, we have $F_{\vec{s}} \in \mathscr{C}(F_{\vec{s}}) = \mathscr{C}(F_{\vec{s}'}) = \mathrm{Orb}(F_{\vec{s}'})$. $\square$

We conclude with the analogue of Item 5 from Theorem 1.15, which states that $\{G, G^{*}\} = \{H \mid H_m = G_m\}$, or, equivalently, $H_m = G_m$ if and only if $H \in \{G, G^{*}\}$.

Theorem 3.10. *Let G be an embedded graph. Then*

$$\mathrm{Orb}(G) = \{H \mid H_m \cong G_m\},$$

i.e.,

$$H_m \cong G_m \iff H \in \mathrm{Orb}(G).$$

Proof. $\mathrm{Orb}(G) = \mathscr{C}(G_m) = \{H \mid H_m \cong G_m\}$, where the first equality follows from Theorem 3.8 and the second from Theorem 3.5. $\square$

3.2 A Structural Hierarchy and Corresponding Dualities

In this section, we construct a hierarchy of graph structures, ranging from abstract graphs, through cyclically ordered graphs (which are defined below), to embedded graphs. We show that when we consider medial graphs up to equivalence under these varying levels of specificity, then there is a corresponding hierarchy of dualities, ranging from twisted duality, through partial duality, to geometric duality.

Again, our starting point is the identity

$$\{G, G^*\} = \{H \mid H_m = G_m\}. \tag{3.3}$$

We ask what form of duality replaces geometric duality if we substitute some other graph equivalence for the equality in the right-hand side of Eq. (3.3). In particular, we want to characterise the sets

$$\{H \mid H_m = G_m\}, \{H \mid H_m \doteq G_m\} \quad \text{and} \quad \{H \mid H_m \cong G_m\}, \tag{3.4}$$

where we have changed our notion of graph equivalence from equivalence as embedded graphs ($=$), to equivalence as abstract graphs ($\cong$), or as cyclically ordered graphs ($\doteq$), which is defined below. These sets will correspond to a set of generalised duals of G, just as equivalence as embedded graphs correspond to geometric duality as in Eq. (3.3).

3.2.1 Forms of Equivalences

As discussed above, we observe a hierarchy of graph structures. At one end of the hierarchy, we have abstract graphs, and at the other end we have embedded graphs. As an intermediary structure, we define cyclically ordered graphs.

Definition 3.11. A *cyclically ordered graph*, or *cog*, consists of an abstract graph (referred to as the underlying abstract graph) together with a cyclic ordering of the half-edges about each vertex.

We consider cogs up to *vertex reversals*, which are the reversals of the cyclic order of the edges about some vertices. This is similar to the local switches of a vertex of a signed rotation system, except that there are no signs on the edges to toggle. Just as ribbon graphs may be thought of as discs with ribbons attached, cogs may be thought of as discs with strings attached.

Definition 3.12. We say that two cogs G and H are *equivalent*, if there is an equivalence of the underlying abstract graphs that preserves or reverses the cyclic orders at the vertices.

Just as an abstract graphs can be obtained from an embedded graph by forgetting some structure, so can cogs.

Definition 3.13. If G is an embedded graph, then its *underlying cog* is the cog that results from the underlying abstract graph of G together with a cyclic order at each vertex induced by a local orientation at that vertex.

Note that the underlying cog is a choice of representative of equivalence class under vertex reversals. While we may associate a canonical underlying cog (i.e., a canonical representative of the equivalence class) for an oriented embedded graph by choosing cyclic orders at each vertex that are consistent with the orientation, this is not the case for unorientable (or unoriented) embedded graphs.

We emphasise that a cyclically ordered graph is not an embedded graph (or a rotation system). As we noted in Sect. 1.1.8, embedded graphs correspond to signed rotation systems, and a cyclically ordered graph has no signs on its edges. For this reason we have abandoned the terminology "combinatorial map" used in [30] in favour of the cyclically ordered graphs used here, as traditional usage of "map" evokes surface embeddings of graphs.

We now use the concept of a cog to define an equivalence of embedded graphs that is stronger than equivalence as abstract graphs, yet weaker than equivalence as embedded graphs.

Definition 3.14. We say that two embedded graphs G and H are *equivalent as cyclically ordered graphs*, and write $G \doteq H$, if G and H have equivalent underlying cogs.

Thus, we have a hierarchy of increasing structure, from an abstract graph, to a cyclically ordered graph, to an embedded graph. Clearly, two equivalent embedded graphs have the same underlying cog and the same underlying abstract graph, while two equivalent cogs have the same underlying abstract graph. Thus, the three notions of equivalence of embedded graphs form a nested sequence of equivalence relations on the set of embedded graphs:

$$G = H \implies G \doteq H \implies G \cong H.$$

Example 3.15. The embedded graphs shown in Fig. 3.9a–c are all embeddings of the same abstract graph (a θ-graph with a loop) and are therefore all equivalent as abstract graphs, but not as embedded graphs. Furthermore, the embedded graphs shown in Fig. 3.9a, c are equivalent as cogs, whereas those shown in Fig. 3.9a, b are not equivalent as cogs.

3.2.2   Induced Dualities

We can now answer Question 3.2, which asks what types of dualities are generated when we change the type of graph equivalence among medial graphs, and what forms of graph equivalence are generated when we change our concept of duality. That is, we determine how the hierarchy of graph equivalences is captured by a hierarchy of dualities.

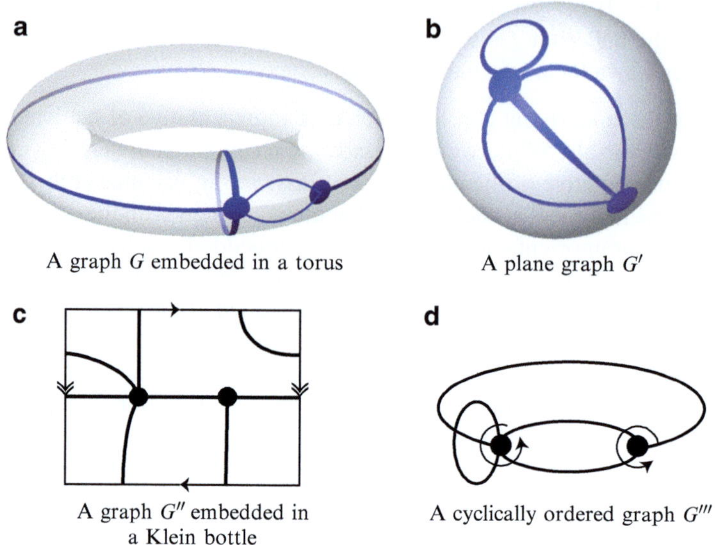

Fig. 3.9 Different embeddings of the same abstract graph

We will see that geometric duality corresponds to equivalence as embedded graphs, partial duality corresponds to equivalence as cogs, and twisted duality corresponds to equivalence as abstract graphs:

1. $\{H \,|\, H_m = G_m\} = \{G, G^*\}$
2. $\{H \,|\, H_m \doteq G_m\} = \mathrm{Orb}_{(\delta)}(G)$
3. $\{H \,|\, H_m \cong G_m\} = \mathrm{Orb}(G)$

Item 1 was stated in Theorem 1.15, and Item 3 is Theorem 3.10. Thus we focus our attention here on Item 2.

Our goal is to show that partial duality is generated by equivalence as cogs, i.e.,

$$\mathrm{Orb}_{(\delta)}(G) = \{H \,|\, H \doteq G_m\}.$$

This will extend Eq. (3.1), and restrict Theorem 3.10, to partial duality. To prove this result we proceed as we did for Theorem 3.10, but now finding results for equivalence as cogs that correspond to those for $=$ given by Theorem 1.15, and to those for $\cong$ given in Sect. 3.1.

We will need the following lemma.

Lemma 3.16.

1. Let G be an embedded graph and $A \subseteq E(G)$. Then $G \doteq G^{\tau(A)}$.
2. Let G and H be embedded graphs such that $G \doteq H$. Then $G = H^{\tau(A)}$ for some $A \subseteq E(H)$.

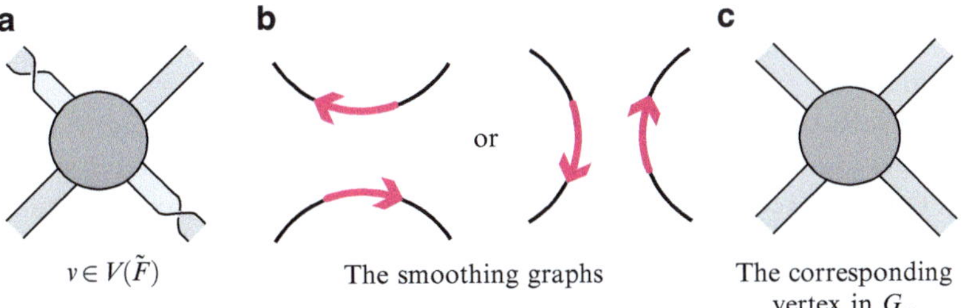

Fig. 3.10   Forming the medial graph of a smoothing graph

Proof. For the first item, view G and $G^{\tau(A)}$ as ribbon graphs and arbitrarily choose a cyclic order of the half-edges incident to each vertex of G. This induces a cyclic order of the half-edges at each vertex of $G^{\tau(A)}$. Equipping the underling abstract graphs of G and $G^{\tau(A)}$ with these cyclic orders results in the same cog. Thus $G \doteq G^{\tau(A)}$.

For the second item, suppose G and H are embedded graphs such that $G \doteq H$, and let v be a vertex in G and $\hat{v}$ be the corresponding vertex in H. Similarly, let e be an edge in G and $\hat{e}$ be the corresponding edge in H. Then, since the underlying abstract graphs of G and H are equivalent, e and $\hat{e}$ can only differ by their numbers of half-twists (i.e., they are related by the operation $\tau(e)$). Furthermore, since $G \doteq H$, if the cyclic orders of the incident half-edges to v and $\hat{v}$ are not equal, then one is the reverse of the other and it can be reordered by adding a half-twist to each of its incident edges as in Fig. 1.4. It then follows that $G = H^{\tau(A)}$ for some A. $\square$

Equation (3.1) relied on the construction of Tait graphs. The more general Theorem 3.10 relied upon the more general construction of cycle family graphs. Theorem 3.22 will involve partial duality which is more general than geometric duality, but more specialised than twisted duality. Accordingly, we need a generalisation of Tait graphs that arises as a special case of cycle family graphs. The set that will play the role of Tait graphs here is the set $\mathscr{C}_{(\delta)}(F)$ of smoothing graphs from Definition 3.4. In analogy with Eq. (3.1) and Theorem 3.10, smoothing graphs completely characterise the medial graphs that are equivalent to a given 4-regular cog under the relation $\doteq$.

Theorem 3.17. *Let F be a 4-regular abstract graph and let $\tilde{F}$ be any embedding of F. Then*

$$\mathscr{C}_{(\delta)}(\tilde{F}) = \{G \mid G_m \doteq F\}.$$

Proof. First suppose that G is a smoothing graph of $\tilde{F}$, i.e., $G \in \mathscr{C}_{(\delta)}(\tilde{F})$. We need to show that $G_m \doteq \tilde{F}$. View $\tilde{F}$ as a ribbon graph and consider a vertex v as shown in Fig. 3.10a. Then forming the smoothing graph results in one of the arrow presentations shown in Fig. 3.10b. Forming the medial graph of either of these arrow presentations then gives the ribbon graph shown in Fig. 3.10c. Note that the adjacency information of G_m and $\tilde{F}$, and the cyclic orders at vertices, is preserved.

All that has (possibly) changed is the number of half-twists in the edges. Thus $G_m = \tilde{F}^{\tau(A)}$, for some set of edges A. By Lemma 3.16 $\tilde{F}^{\tau(A)} \doteq \tilde{F}$, so that $G_m \doteq \tilde{F}$, as required.

Conversely, suppose that $G_m \doteq \tilde{F}$. Then, by Lemma 3.16, G_m and $\tilde{F}$ are partial Petrials. As G_m is a medial graph, G can be recovered from it as a Tait graph. This means G can be recovered from G_m by choosing one of the arrow marked vertex states of Fig. 3.2b or c at each vertex. Choosing the same arrow marked vertex state at the corresponding vertices of $\tilde{F}$ then also results in G (as G_m and $\tilde{F}$ differ only in the number of half-twists on their edges). Thus G is a smoothing graph of $\tilde{F}$, completing the proof. $\square$

We get the analogue of Item 2 of Theorem 1.15 (that $(F_{bl})_m = (F_{wh})_m = F$) as an immediate corollary.

Corollary 3.18. *Let F be a 4-regular embedded graph and $\vec{s}$ be a duality state of F. Then $(F_{\vec{s}})_m$ and F are equivalent as cogs, i.e.,*

$$(F_{\vec{s}})_m \doteq F.$$

Proof. If $\vec{s}$ is a duality state, $F_{\vec{s}} \in \mathscr{C}_{(\delta)}(F)$, but by Theorem 3.17, $\mathscr{C}_{(\delta)}(F) = \{G \,|\, G_m \doteq F\}$, so $(F_{\vec{s}})_m \doteq F$. $\square$

The following theorem extends the identity $\{G, G^*\} = \{(G_m)_{bl}, (G_m)_{wh}\}$ to partial duality.

Theorem 3.19. *Let G be an embedded graph. Then the smoothing graphs of its medial graph G_m are exactly its partial duals, i.e.,*

$$\mathscr{C}_{(\delta)}(G_m) = \mathrm{Orb}_{(\delta)}(G).$$

Proof. Let $H = G^{\delta(A)}$. Then the arrow presentation for each edge of H is one of those shown in Fig. 2.4. But these are exactly the arrow presentations of the smoothing graphs of G_m that arise by replacing each vertex of G_m with one of the arrow marked states shown in Fig. 3.2b, c. Thus $H \in \mathscr{C}_{(\delta)}(G_m)$.

Conversely, if $H \in \mathscr{C}_{(\delta)}(G_m)$, then it is obtained by replacing each vertex of G_m with one of the arrow marked states shown in Fig. 3.2b, c. But these are exactly the arrow presentations of the twisted duals of G as shown in Fig. 2.4. Thus $H = G^{\delta(A)}$, completing the proof. $\square$

Similarly, Theorem 1.15 states that $G_m = H_m$ if and only if $H \in \{G, G^*\}$, and Theorem 3.10 has the corresponding result for twisted duality. We can now give the corresponding result for partial duals and cogs.

Corollary 3.20. *Let G and H be embedded graphs. Then*

$$G_m \doteq H_m \iff H = G^{\delta(A)}$$

for some $A \subseteq E(G)$.

Also, we can obtain an analogue of Theorem 1.15, Item 4, which states that $F_{bl} = (F_{wh})^*$.

Corollary 3.21. *If F is a 4-regular embedded graph and $F_{\vec{s}}$ and $F_{\vec{s}'}$ are two smoothing graphs of F, then $F_{\vec{s}}$ and $F_{\vec{s}'}$ are partial duals, i.e.,*

$$F_{\vec{s}} = (F_{\vec{s}'})^{\delta(A)}$$

for some $A \subseteq E(G)$.

Proof. By Theorem 3.17, $F_{\vec{s}}, F_{\vec{s}'} \in \mathscr{C}_{(\delta)}(F) = \{G \,|\, G_m \doteq F\}$. Thus, $(F_{\vec{s}})_m \doteq F \doteq (F_{\vec{s}'})_m$. Then, also by Theorem 3.17, $\mathscr{C}_{(\delta)}(F_{\vec{s}}) = \mathscr{C}_{(\delta)}(F_{\vec{s}'})$. Finally, $F_{\vec{s}} \in \mathscr{C}_{(\delta)}(F_{\vec{s}}) = \mathscr{C}_{(\delta)}(F_{\vec{s}'}) = \mathrm{Orb}_{(\delta)}(F_{\vec{s}'})$. $\square$

We are now able to prove the desired analogues of Eq. (3.1) and Theorem 3.10 in Theorem 3.22, thus showing that equivalence as cogs is the type of embedded graph equivalence that corresponds to partial duality.

Theorem 3.22. *Let G be an embedded graph. Then*

$$\mathrm{Orb}_{(\delta)}(G) = \{H \,|\, H \doteq G_m\}.$$

Proof. We have $\mathrm{Orb}_{(\delta)}(G) = \mathscr{C}_{(\delta)}(G_m) = \{G^{\delta(A)} \,|\, A \subseteq E(G)\}$, where the first equality is by Theorem 3.19 and the second is by Theorem 3.17. $\square$

Let us recap the main ideas in this section. We began with the classical relation between medial graphs and geometric duality: $\{G, G^*\} = \{H \mid H_m = G_m\}$. This led us to the principle that duality and embedded graph equivalence are equivalent concepts. That is, the type of equivalence of medial graphs leads to a kind of duality and vice versa. Moreover, this equivalence of types is realised through the idea of a Tait graph. We determined the duality that corresponds with equivalence as abstract graphs, $\cong$, which was twisted duality, and the duality that corresponds with equivalence as cogs, $\doteq$, which was partial duality. Thus we have given a complete characterisation of the orbits of the full ribbon group action and of its subgroups generated by $\{(\delta, 1, \ldots, 1), (1, \delta, 1, \ldots, 1), \ldots, (1, 1, \ldots, \delta)\}$ and by $\{\delta \ldots \delta\}$. That is, we have shown the following result.

Theorem 3.23. *Let G and H be embedded graphs. Then there is the following hierarchy of graph equivalences and notions of duality:*

1. *G_m and H_m are equivalent as embedded graphs if and only if G and H are geometric duals.*
2. *G_m and H_m are equivalent as cogs (or are partial Petrials) if and only if G and H are partial duals.*
3. *G_m and H_m are equivalent as abstract graphs if and only if G and H are twisted duals.*

Table 3.2 summarises the extensions of the basic properties of Tait graphs, medial graphs, and equivalence given in Theorem 1.15.

Table 3.2 A summary of the connections between equivalences and dualities

Geometric duality	Partial duality	Twisted duality
$\{F_{bl}, F_{wh}\} = \{G \mid G_m = F\}$	$\mathscr{C}_{(\delta)}(F) = \{G \mid G_m \doteq F\}$	$\mathscr{C}(F) = \{G \mid G_m \cong F\}$
$(F_{bl})_m = (F_{wh})_m = F$	$(F_{\vec{s}})_m \doteq F$	$(F_{\vec{s}})_m \cong F$
	$F_{\vec{s}}$ a smoothing graph	$F_{\vec{s}}$ a cycle family graph
$\{(G_m)_{bl}, (G_m)_{wh}\} = \{G, G^*\}$	$\mathscr{C}_{(\delta)}(G_m) = \mathrm{Orb}_{(\delta)}(G)$	$\mathscr{C}(G_m) = \mathrm{Orb}(G)$
$F_{bl} = (F_{wh})^*$	$F_{\vec{s}} = (F_{\vec{s}'})^{\delta(A)}$	$F_{\vec{s}} = F_{\vec{s}'}^{\Gamma}$
	$F_{\vec{s}}, F_{\vec{s}'}$ smoothing graphs	$F_{\vec{s}}, F_{\vec{s}'}$ cycle family graphs
$\{G, G^*\} = \{H \mid H_m = G_m\}$	$\mathrm{Orb}_{(\delta)}(G) = \{H \mid H \doteq G_m\}$	$\mathrm{Orb}(G) = \{H \mid H_m \cong G_m\}$

The preceding results immediately beg the question of how to characterise the many other orbits of the ribbon group action in terms of embedded graph equivalence. Some preliminary results in this direction were given in [30]; however, much work remains to be done in this area.

3.3   Properties of Some Special Orbits

We use the results of Sects. 3.1.2 and 3.2, the relationships among partial duals, twisted duals and medial graphs, and a result of Las Vergnas [65] and Kotzig [62], to deduce some properties of the orbits under the ribbon group action discussed in Sect. 2.4.2.

Proposition 3.24. *Let G be a plane graph. Then*

$$\max\{v(H) \mid H \in \mathrm{Orb}(G)\} = \max\{v(H) \mid H \in \mathrm{Orb}_{(\delta)}(G)\}.$$

Proof. Las Vergnas' Proposition 6.1 from [65] implies that the maximum number of circuits in any duality state of G_m is equal to the maximum number of circuits in any state of G_m. Since the cycles in the states of G_m form the vertices of the cycle family graphs, we have

$$\max\{v(H) \mid H \in \mathscr{C}_{(\delta)}(G_m)\} = \max\{v(H) \mid H \in \mathscr{C}(G_m)\}.$$

The result then follows since $\mathscr{C}_{(\delta)}(G_m) = \mathrm{Orb}_{(\delta)}(G)$, by Theorem 3.17 and Corollary 3.20; and since $\mathscr{C}(G_m) = \mathrm{Orb}(G)$, by Theorem 3.10. $\square$

The following corollary relates the number of spanning trees of a ribbon graph G and of its dual G^* to the number of *bouquets* (i.e., embedded graphs with exactly one vertex) in $\mathrm{Orb}_{(\delta)}(G)$ and to the Tutte polynomial, which is discussed further in Chap. 4.

Proposition 3.25. *Let G be a graph embedded in the plane, the torus, or the real projective plane. In addition, let $\mathscr{B}_{(\delta)}(G)$ denote the number of bouquets in*

$\mathrm{Orb}_{(\delta)}(G)$. *Then $\mathscr{B}_{(\delta)}(G)$ is bounded above by the total number of spanning trees in G and G^*. Moreover,*

$$\mathscr{B}_{(\delta)}(G) \leq 2T(G;1,1),$$

where $T(G;x,y)$ is the Tutte polynomial of G.

Proof. By Las Vergnas' Corollary 2.4 from [65], every duality state of G_m that contains exactly one cycle corresponds to a unique spanning tree in G or G^*. (The plane case of this result is due to Kotzig [62].) Moreover, every duality state of G_m that contains exactly one cycle gives rise to a (not necessarily distinct) cycle family graph in $\mathscr{C}_{(\delta)}(G_m)$ that has exactly one vertex. Therefore,

$$|\{\text{spanning trees of } G \text{ or } G^*\}| \geq |\{H \in \mathscr{C}_{(\delta)}(G_m) \mid v(H) = 1\}|.$$

Since $\mathscr{C}_{(\delta)}(G_m) = \mathrm{Orb}_{(\delta)}(G)$, by Theorem 3.19, it follows that

$$|\{\text{spanning trees of } G \text{ or } G^*\}| \geq |\{H \in \mathrm{Orb}_{(\delta)}(G) \mid v(H) = 1\}| = \mathscr{B}_{(\delta)}(G).$$

The result then follows by noting that the number of spanning trees in a connected graph G is $T(G;1,1)$, and that $T(G,x,y) = T(G^*;x,y)$, and so

$$2T(G;1,1) = T(G;1,1) + T(G^*;1,1) = |\{\text{spanning trees of } G \text{ or } G^*\}| \geq \mathscr{B}_{(\delta)}(G).$$

$$\square$$

The above result can also be extended to all embedded graphs by using quasi-trees. A *quasi-tree* is an embedded graph with exactly one boundary component (or face).

Corollary 3.26. *Let G be an embedded graph, then the number of bouquets in $\mathrm{Orb}_{(\delta)}(G)$ is bounded above by the number of spanning quasi-trees of G.*

Proof. From [78], if $A \subseteq E(G)$, and $A^c = E(G)\backslash A$, then the number of vertices of $G^{\delta(A)}$ is equal to the number of boundary components of $G - A^c$. It then follows that $G^{\delta(A)}$ is a bouquet if and only if $G - A^c$ is a spanning quasi-tree. The result then follows, noting that the partial duals need not be distinct. $\square$

Similar results hold for the orbit of an embedded graph under the action of the subgroup of the ribbon group generated by half-twists.

Proposition 3.27. *Let G be an embedded graph. Then*

1. *$|\mathrm{Orb}_{(\tau)}(G)|$ is bounded above by two raised to the power of the number of cycles in G.*
2. *If G is bipartite, then $G^{\tau(E(G))} = G$. Furthermore (Wilson [100]) if G is an orientable regular map, then $G^{\tau(E(G))} = G$ if and only is G is bipartite.*

Proof. We will work in the language of ribbon graphs. Both results follow from the observations that $\tau(e)$ changes the ribbon graph by adding a half-twist to the edge e, so up to the equivalence of ribbon graphs, $\tau(A)$ can only act by adding or removing half-twists to cycles, for each $A \subseteq E(G)$. The first result then follows by observing that $\tau(A)$ can only act by changing the orientability of a set of cycles of G.

That all embeddings of bipartite graphs are self-Petrial follows because every cycle is even, so its orientability is unchanged if every edge is given a half-twist and the cyclic order of half-edges incident to each vertex is unchanged. The "if and only if" statement is due to Wilson [100] and does not in general hold for unorientable embedded graphs. For example, the hemi-dodecahedron in the real projective plane is self-Petrial, but not bipartite. □

Chapter 4
Interactions with Graph Polynomials

In this chapter, after a brief review of some classical graph polynomials as well as deletion and contraction reductions, we turn to polynomials of embedded graphs, beginning with the topological transition polynomial of [30]. The topological transition polynomial interacts with twisted duality in a particularly natural way, leading to a generalised duality identity, and a three-term deletion-contraction relation. The topological transition polynomial specialises to the topological Penrose polynomial and agrees with various forms of the topochromatic polynomial, and hence the ribbon graph polynomial of Bollobás and Riordan. Thus, the identities for the topological transition polynomial lead in turn to new results for these polynomials, including reformulations of the Four Colour Theorem. We survey these results in this chapter. Later, in Chap. 5, we will survey recent connections between the topological graph polynomials discussed here (in particular the topochromatic polynomial and hence both signed and unsigned versions of the ribbon graph polynomial of Bollobás and Riordan) and knot polynomials.

4.1 Classical Graph Polynomials

First, we briefly review the classical Tutte polynomial, and mention the few of its many properties that will concern us here. Comprehensive surveys of the Tutte polynomial may be found in [4, 10, 11, 26, 27, 83].

Definition 4.1. If $G = (V, E)$ is an abstract graph, then the *Tutte polynomial* of G is given by the following recursion: if e is neither a loop nor a bridge, then

$$T(G; x, y) = T(G - e; x, y) + T(G/e; x, y); \tag{4.1}$$

and if G consists of i bridges and j loops, then

$$T(G; x, y) = x^i y^j. \tag{4.2}$$

J.A. Ellis-Monaghan and I. Moffatt, *Graphs on Surfaces: Dualities, Polynomials, and Knots*, SpringerBriefs in Mathematics, DOI 10.1007/978-1-4614-6971-1_4, © Joanna A. Ellis-Monaghan, Iain Moffatt 2013

Thus, the Tutte polynomial may be calculated by specifying an ordering of the edges and repeatedly applying Eq. (4.1) until only bridges and loops remain, and then applying Eq. (4.2). The Tutte polynomial is well defined in that the result of this recursive process is independent of the ordering of the edges. There are several ways to prove this. One way is to show, by induction, that it is equal to the following rank-nullity, or state sum, formulation, which is not dependent on any ordering of the edges:

$$T(G;x,y) = \sum_{A \subseteq E(G)} (x-1)^{r(E)-r(A)}(y-1)^{n(A)}. \tag{4.3}$$

From this form, it is relatively easy to see that Tutte polynomial is equal, up to a prefactor, to the *dichromatic polynomial*

$$Z_T(G;u,v) = \sum_{A \subseteq E(G)} u^{k(A)}v^{|A|},$$

specifically,

$$u^{k(G)}v^{r(G)}T\left(G;\frac{u+v}{v},v+1\right) = Z_T(G;u,v). \tag{4.4}$$

The classical Tutte polynomial has the elegant duality property that the Tutte polynomial of a plane graph G is the same as that of its geometric dual G^* with the roles of the variables x and y permuted:

$$T(G;x,y) = T(G^*;y,x). \tag{4.5}$$

In this chapter, we will be particularly interested in analogous twisted-duality identities for various topological graph polynomials.

We are also interested in colouring information encoded by topological graph polynomials. We recall here that the classical Tutte polynomial also encodes colouring information through its relation to the chromatic polynomial.

Definition 4.2. The *chromatic polynomial*, $\chi(G;x) \in \mathbb{Z}[x]$, is defined recursively by

$$\chi(G;x) = \chi(G-e;x) - \chi(G/e;x)$$

and

$$\chi(E_n;x) = x^n,$$

where E_n is the edgeless graph on n vertices.

An *n-colouring* ϕ of G is a mapping $\phi : V(G) \rightarrow \{1,2,\ldots,n\}$. (That is, an n-colouring is an assignment of colours from the set $\{1,2,\ldots,n\}$ to the vertices of G.) An n-colouring ϕ is said to be *proper* if $\phi(u) \neq \phi(v)$ whenever $u,v \in V(G)$ are adjacent. The chromatic polynomial can be shown to count proper n-colourings:

Proposition 4.3. *Let G be a graph, $\chi(G;x)$ be its chromatic polynomial, and $n \in \mathbb{N}$. Then $\chi(G;n)$ is the number of proper n-colourings of G.*

It can be shown that the chromatic polynomial can be obtained as an evaluation of the Tutte polynomial:

$$\chi(G;x) = x^{k(G)}(-1)^{r(G)}T(G;1-x,0).$$

4.2 Deletion, Contraction, and Medial Graphs

Before moving on to polynomials of embedded graphs, we examine the basic operations of deletion and contraction for embedded graphs, and especially how they interact with medial graphs.

Deletion of the edge of a ribbon graph is defined much as for abstract graphs. Let G be a ribbon graph and e be an edge of G. Then $G-e$ denotes the ribbon graph obtained from G by removing, or *deleting*, the edge e. Edge deletion in terms of arrow presentations, ram graphs, and signed rotation systems can similarly be defined in a straightforward manner. In terms of arrow presentations, the two e-labelled arrows are deleted. In terms of ram graphs, delete either the two e-labelled arrows or the edge e. For signed rotation systems, delete the edge e and remove it from the cyclic orderings at its incident vertices.

Care is needed when deleting edges of cellularly embedded graphs or band decompositions. For example, if G is a cellularly embedded graph and e is a bridge of G, then simply removing the edge e from G will result in a graph that is no longer cellularly embedded. Similar issues arise when a bridge is removed from a band decomposition. Accordingly, deletion for cellularly embedded graphs and band decompositions is often best done by converting to the language of ribbon graphs, carrying out the deletion, then converting back to the language of cellularly embedded graphs or band decompositions. This is how we define deletion of an edge for these representations of embedded graphs.

Contracting non-loop edges of an embedded graph is fairly straightforward. Let G be a ribbon graph and e be a non-loop edge of G that has incident vertices u and v. Then e, u, and v are all discs, and their union $e \cup u \cup v$ is also a disc. We can then define G/e to be the ribbon graph obtained from G by replacing e, u, and v with a single vertex $e \cup u \cup v$. We can define contraction for non-loop edges of other realisations of embedded graphs by converting to the language of ribbon graphs, carrying out the contraction, then converting back to the original language. In particular, if $G \subset \Sigma$ is a cellularly embedded graph with non-loop edge e, then G/e is obtained by identifying all of the points on e (i.e., by forming the quotient space Σ/e).

Additional care needs to be taken when contracting loops of an embedded graph. To see why, consider what happens when we try to lift the above definition of contraction from non-loop edges to loops. If G is a ribbon graph and e is an orientable loop, then $e \cup u \cup v$ is not a disc so the definition of contraction for non-loop edges of ribbon graphs above does not extend. However, if e is a non-orientable loop in a ribbon graph, then $e \cup u \cup v$ is a disc, and we can define G/e just as in the

Type of edge	G	G/e
Not a loop	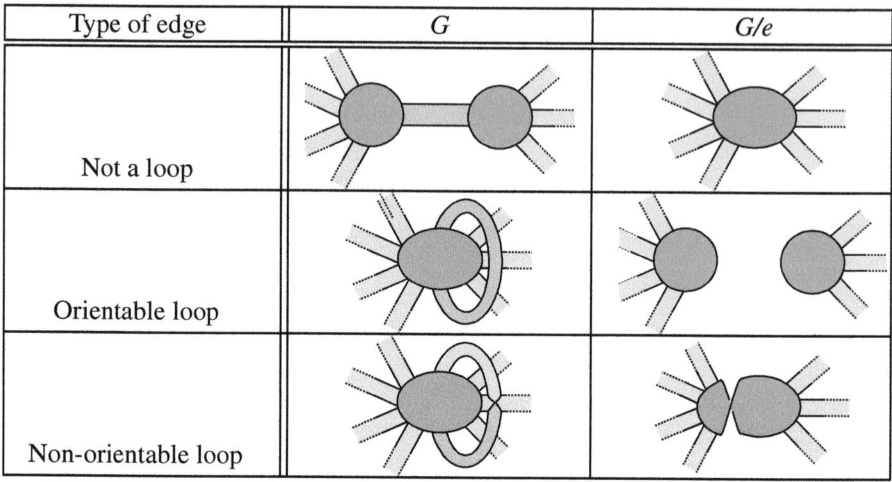	
Orientable loop		
Non-orientable loop		

Fig. 4.1 Contracting an edge of a ribbon graph

non-loop case. On the other hand, if G is a cellularly embedded graph and e is a loop, then Σ/e can be a surface or a pinched surface, so again there are difficulties when defining contraction for cellularly embedded graphs.

To extend contraction to loops, consider the definition of contraction of a non-loop edge in terms of arrow presentations. Let G be an arrow presentation and e be a non-loop edge, then G/e is obtained as follows. Suppose α and β are the two e-labelled arrows. Connect the tip of α to the tail of β with a line segment; and connect the tip of β to the tail of α with another line segment. Delete α, β, and the arcs of the circles on which they lie. This results in an arrow presentation for G/e. Observe (1) that we did not use the fact that e was not a loop, and (2) by comparing this construction with Definition 2.3, we see that we have obtained $G^{\delta(e)} - e$. These considerations lead to the following definition of contraction of an edge of an embedded graph that works for all edge types (both loops and non-loops) and agrees with all of the constructions of G/e discussed above.

Definition 4.4. Let G be an embedded graph and e be any edge of G. Then the embedded graph G/e obtained from G by *contracting* the edge e is defined by

$$G/e := G^{\delta(e)} - e.$$

Figure 4.1 shows the local effect of contracting an edge of a ribbon graph.

We note that edge contraction as defined in Definition 4.4 was considered by Bollobás and Riordan in Sect. 7 of [6] (where the language of partial duals was not used) and was observed to agree with $G^{\delta(e)} - e$ by Chmutov in [16].

We conclude by setting up some additional notation. If G is an embedded graph and $A \subseteq E(G)$, then $G - A$ is the embedded graph obtained from G by deleting all of the edges in A and G/A is the embedded graph obtained by contracting all of the edges in A.

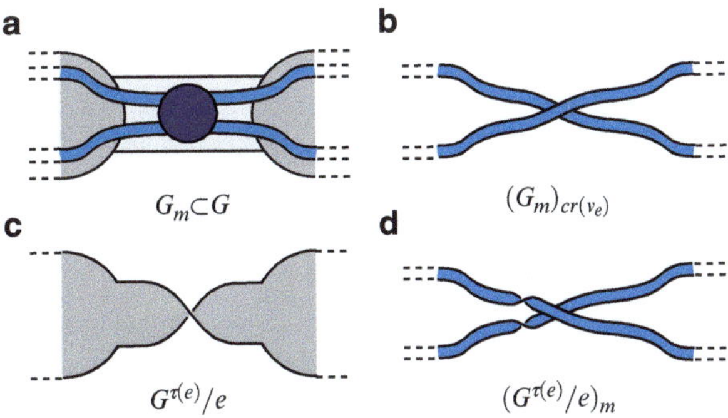

Fig. 4.2 Proving that $(G_m)_{cr(v_e)}$ and $(G^{\tau(e)}/e)_m$ are partial Petrials

Deletion and contraction are reflected in the medial graph. For Proposition 4.5 we use the notation for vertex states from Sect. 1.5.2.

Proposition 4.5. *Let G be an embedded graph with embedded, canonically checkerboard coloured medial graph G_m, and let e be any edge of G, with v_e the associated vertex in G_m. Then*

1. $(G_m)_{bl(v_e)} = (G-e)_m$.
2. $(G_m)_{wh(v_e)} = (G/e)_m$.
3. $(G_m)_{cr(v_e)}$ *and* $(G^{\tau(e)}/e)_m$ *are partial Petrials.*

Proof. The result can be seen by considering the medial graph G_m as a ribbon graph embedded in the ribbon graph G and doing similarly for $(G-e)_m$, $(G/e)_m$, and $(G^{\tau(e)}/e)_m$. We will illustrate the proof by showing that $(G_m)_{cr(v)_e}$ and $(G^{\tau(e)}/e)_m$ are partial Petrials in the case when e is a non-loop edge, leaving the remaining cases for the reader. Figure 4.2a shows the medial graph G_m embedded in the ribbon graph G, and Fig. 4.2b shows $(G_m)_{cr(v)_e}$. On the other hand, $G^{\tau(e)}/e$ is shown in Fig. 4.2c and its medial graph in Fig. 4.2d. It is readily seen that $(G_m)_{cr(v)_e}$ and $(G^{\tau(e)}/e)_m$ are partial Petrials, as required. $\square$

4.3 Twisted Duals and the Topological Transition Polynomial

The generalised transition polynomial, $q(G;W,t)$, of [32] is a multivariate graph polynomial that generalises Jaeger's transition polynomial [51]. The transition polynomial assimilates the Penrose polynomial (see Sect. 4.4.1) and Kauffman bracket (see Sect. 5.4.1) and agrees with the Tutte polynomial via a medial graph construction (see Sect. 4.5). The generalised transition polynomial was adapted to embedded graphs in [30], resulting in the topological transition polynomial, which relates analogously to versions of the Penrose polynomial, Kauffman bracket, and Tutte polynomial that have likewise been extended to graphs and knot diagrams in surfaces. Here we review the topological transition polynomial and its interaction with the ribbon group action.

Although $q(G;W,t)$ can be defined for arbitrary Eulerian graphs (see [32] or [34] for details), here we will restrict q to 4-regular embedded graphs (typically medial graphs) as this is the level of generality needed for our current application.

4.3.1 The Topological Transition Polynomial

A *weight system*, $W(F)$, of any 4-regular graph F (embedded or not) is an assignment of a weight in a unitary ring $\mathscr{R}$ to every vertex state of F. (We simply write W for $W(F)$ when the graph is clear from context.) If s is a state of F, then the *state weight* of s is $\omega(s) := \prod_{v \in V(F)} \omega(v,s)$, where $\omega(v,s)$ is the vertex state weight of the vertex state at v in the graph state s. Recall that a graph state s consists of a set of disjoint closed curves, which we refer to as the components of the state, denoting the number of them by $c(s)$.

Definition 4.6. Let F be a 4-regular graph having weight system W with values in a unitary ring $\mathscr{R}$. Then the state model formulation of the *generalised transition polynomial* is

$$q(F;W,t) = \sum_s \omega(s)t^{c(s)},$$

where the sum is over all graph states s of F.

The generalised transition polynomial is easily seen to be multiplicative on disjoint unions. Also, it may be computed recursively by, for one vertex at a time, taking all possible vertex states and forming a weighted sum of the resulting graphs with the weights given by the state weights. This eventually leads to a weighted sum of sets of closed curves. Each closed curve is then evaluated to t. Details of this can be found in [32, 34].

We now restrict our attention further to embedded medial graphs and particular weight systems determined by the embeddings. Because of these restrictions, we will call the generalised transition polynomial specialised for this application the *topological transition polynomial* and define it as follows.

Definition 4.7. Let G be an embedded graph with embedded medial graph G_m. Define the *medial weight system*, $W_m(G_m)$, using the canonical checkerboard colouring of G_m as follows. A vertex v has state weights given by an ordered triple $(\alpha_v, \beta_v, \gamma_v)$, indicating the weights of the white smoothing state, black smoothing state, and crossing state, in that order. We write (α, β, γ) for the set of these ordered triples, indexed equivalently either by the vertices of G_m or by the edges of G. (In the latter case we refer to (α, β, γ) as a weight system for G.) Then the *topological transition polynomial* of G is

$$Q(G,(\alpha,\beta,\gamma),t) := q(G_m;W_m,t).$$

Since the generalised transition polynomial can be computed recursively, so can the topological transition polynomial.

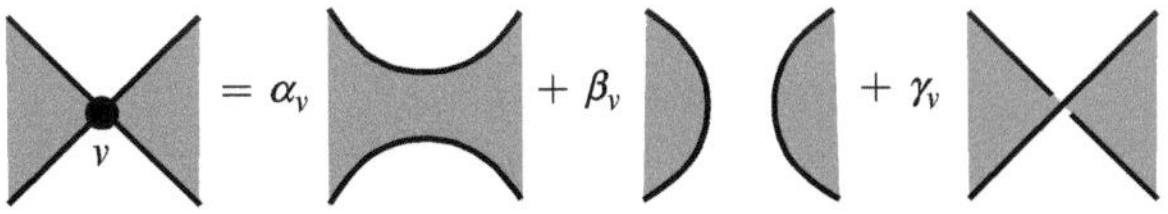

Fig. 4.3 The linear recursion relation of the topological transition polynomial

Proposition 4.8. *The topological transition polynomial may be computed by repeatedly applying the following linear recursion relation at each $v \in V(G_m)$, and, when there are no more vertices of degree 4 to apply it to, evaluating each of the resulting closed curves to an independent variable t:*

$$q(G_m, W_m, t) = \alpha_v q((G_m)_{wh(v)}, W_m, t) + \beta_v q((G_m)_{bl(v)}, W_m, t) + \gamma_v q((G_m)_{cr(v)}, W_m, t).$$

(A pictorial representation of this relation is given in Fig. 4.3.)

Example 4.9. For example, if $G =$ ●—u—●—v—● , then $G_m =$ (figure), and consequently

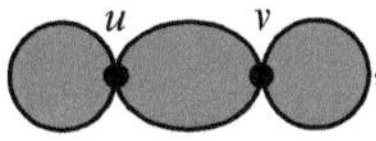

$$Q(G; (\boldsymbol{\alpha}, \boldsymbol{\beta}, \boldsymbol{\gamma}), t)$$

$$= \alpha_u \,(\text{figure})\, + \beta_u \,(\text{figure})\, + \gamma_u \,(\text{figure})$$

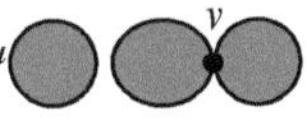

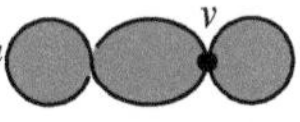

$$= \alpha_u (\alpha_v \,(\text{figure})\, + \beta_v \,(\text{figure})\, + \gamma_v \,(\text{figure})\,) + \cdots$$

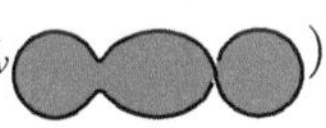

$$= \alpha_u \alpha_v t + \alpha_u \beta_v t^2 + \alpha_u \gamma_v t + \cdots$$

The topological transition polynomial has a deletion-contraction reduction, with an extra "twist-contract" term. In subsequent sections we will see that this property leads to deletion-contraction reductions for other polynomials that are related to it.

We emphasise that in the following theorem the contraction of a loop e (or any other edge e for that matter) is defined by $G/e := G^{\delta(e)} - e$.

Theorem 4.10. *Let G be an embedded graph and $e \in E(G)$. Then*

$$Q(G; (\boldsymbol{\alpha}, \boldsymbol{\beta}, \boldsymbol{\gamma}), t) = \alpha_e Q(G/e; (\boldsymbol{\alpha}, \boldsymbol{\beta}, \boldsymbol{\gamma}), t)$$

$$+ \beta_e Q(G - e; (\boldsymbol{\alpha}, \boldsymbol{\beta}, \boldsymbol{\gamma}), t) + \gamma_e Q(G^{\tau(e)}/e; (\boldsymbol{\alpha}, \boldsymbol{\beta}, \boldsymbol{\gamma}), t),$$

where, on the right-hand side, $(\boldsymbol{\alpha}, \boldsymbol{\beta}, \boldsymbol{\gamma})$ denotes the weight system for G restricted to G/e, $G - e$, or $G^{\tau(e)}/e$ which is obtained by eliminating the weights for e.

Proof. This identity follows from Definition 4.7 and Propositions 4.5 and 4.8 upon observing that twisting an edge of a medial graph does not change the number of components in a transition state. $\qquad\square$

4.3.2 The Topological Transition Polynomial and the Ribbon Group Action

In [34] it was shown that the topological transition polynomial has the duality relation

$$q(G_m; W_m, t) = q(G_m^*; W_m^*, t),\qquad(4.6)$$

or equivalently,

$$Q(G;(\boldsymbol{\alpha},\boldsymbol{\beta},\boldsymbol{\gamma}),t) = Q(G^*;(\boldsymbol{\beta},\boldsymbol{\alpha},\boldsymbol{\gamma}),t),\qquad(4.7)$$

where G^* is the geometric dual of G and W_m^* is the weight system that derives from exchanging the order of α_v and β_v in the weight system at each vertex. This led to a new duality result for the topological Tutte polynomial in [34] (see Sect. 4.5 and Eq. (4.45)), extending the classical relation $T(G;x,y) = T(G^*;y,x)$ for the Tutte polynomial of plane graphs. We will now see that the above duality relation for the topological transition polynomial is in fact a special case of a more general twisted duality relation. This twisted duality relation says that the topological transition polynomial of G is the same as that of any twisted dual of G, provided the weight system is appropriately permuted. We will also see that Eqs. (4.6) and (4.7) follow as a special case.

The group $\mathfrak{G} = \langle \delta, \tau \mid \delta^2, \tau^2, (\tau\delta)^3 \rangle$ is isomorphic to S_3 via $\eta : \delta \mapsto (1\,2)$ and $\eta : \tau \mapsto (1\,3)$.

Furthermore, the symmetric group S_3 acts on the ordered triple of the weight system at a vertex by permutation. That is, if $\xi \in \mathfrak{G}$, then $\eta(\xi)$ acts on $(\alpha_{v_e}, \beta_{v_e}, \gamma_{v_e})$ by permuting the entries. This action by S_3 on the state weights at a vertex can be extended to an action of S_3^n on the vertex state weights of graphs with n linearly ordered vertices. We will not formally define this action here and instead define the order independent analogue of the action, which is more convenient for our applications. This action allows us to use the ribbon group to modify the medial weight system of an embedded medial graph.

Definition 4.11. Let G_m be the canonically checkerboard coloured embedded medial graph of an embedded graph G, with the vertices of G_m indexed by the edges of G, and medial weight system W_m (or equivalently $(\boldsymbol{\alpha},\boldsymbol{\beta},\boldsymbol{\gamma})$). Using Notation 2.19, let $\Gamma = \prod_{i=1}^{6} \xi_i(A_i)$ where the A_i's partition $E(G)$ and the ξ_i's are the six elements of $\mathfrak{G}$. Then $W_m{}^{\Gamma}$ (or $(\boldsymbol{\alpha},\boldsymbol{\beta},\boldsymbol{\gamma})^{\Gamma}$), the *weight system permuted by* Γ, has the ordered triple of the weight system at a vertex v_e given by $\eta(\xi_i)(\alpha_{v_e}, \beta_{v_e}, \gamma_{v_e})$ when $e \in A_i$. Here $\eta(\xi_i)$ acts on $(\alpha_{v_e}, \beta_{v_e}, \gamma_{v_e})$ by permuting the entries by $\eta(\xi_i)$.

Theorem 4.12. *Let G be an embedded graph with embedded medial graph G_m, and let $\Gamma = \prod_{i=1}^{6} \xi_i(A_i)$ where the A_i's partition $E(G)$ and the ξ_i's are the six elements of $\mathfrak{G}$. Then,*

$$q(G_m; W_m, t) = q\left(G_m^{\Gamma}; W_m^{\Gamma}, t\right),$$

or equivalently,

$$Q(G;(\boldsymbol{\alpha},\boldsymbol{\beta},\boldsymbol{\gamma}),t) = Q(G^{\Gamma},(\boldsymbol{\alpha},\boldsymbol{\beta},\boldsymbol{\gamma})^{\Gamma},t).$$

Table 4.1 Figures used in the proof of Theorem 4.12

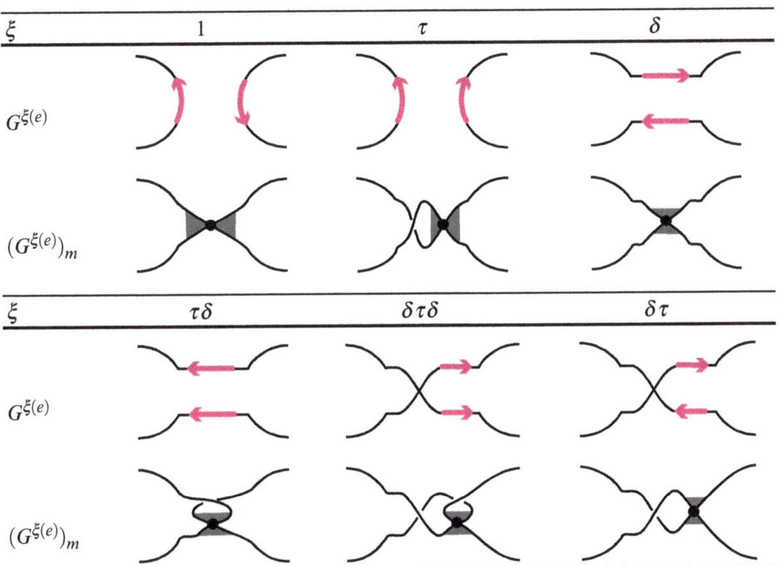

Proof. It suffices to prove this for a single edge e of G, i.e., to prove the property for G and $G^{\xi(e)}$ where $\xi \in \mathfrak{G}$. Consider an edge e of G represented as an arrow presentation. Table 4.1 shows an arrow presentation for $G^{\xi(e)}$, for each $\xi \in \mathfrak{G}$, locally at the e-labelled arrows. Note that G is shown as the $G^{1(e)}$ entry.

In comparing $Q(G^{\xi(e)}, (\alpha, \beta, \gamma)^{\xi(e)}, t)$, for different values of ξ, we can assume that the arrow presentations for $G^{\xi(e)}$ are identical outside of a local region containing the two e-labelled arrows and that these local regions are as shown in Table 4.1. Then, all that matters in the computations for $Q(G^{\xi(e)}, (\alpha, \beta, \gamma)^{\xi(e)}, t)$ is how the medial graphs $(G^{\xi(e)})_m$ differ in this local region and where the checkerboard colouring is. This is also shown in Table 4.1 in the $(G^{\xi(e)})_m$ labelled rows.

All that remains to prove the theorem is to compare the contributions of the vertex v_e to the calculations of $Q(G; (\alpha, \beta, \gamma), t)$ and $Q(G^{\xi(e)}, (\alpha, \beta, \gamma)^{\xi(e)}, t)$, for each ξ. Figure 4.4 shows the comparison for $\xi = \delta\tau\delta$. The figure shows the linear recursion of Fig. 4.3 applied to the vertex v_e of G and of $G^{\delta\tau\delta(e)}$ using the weight systems (α, β, γ) and $(\alpha, \beta, \gamma)^{\delta\tau\delta(e)}$, respectively. It is readily seen that the contributions of the vertices to the topological transition polynomials are equal. Similar computations for the remaining elements of $\mathfrak{G}$, which are left as an exercise, complete the proof of the theorem.   □

Observe that by taking $\Gamma = \delta(E(G))$, the result in Eq. (4.6) is now just an immediate corollary of Theorem 4.12.

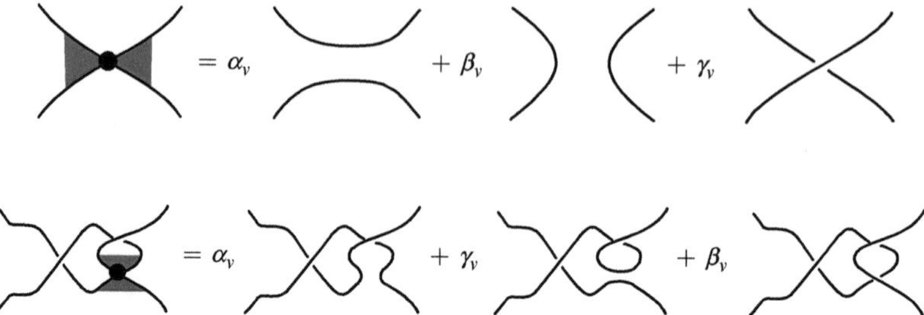

Fig. 4.4 A figure used in the proof of Theorem 4.12

If we apply an element of the ribbon group to a single edge, then we have the following recursion relation with appropriately adjusted weights.

Corollary 4.13. *Let G be an embedded graph, $e \in E(G)$, and $\xi \in \mathfrak{G}$. Further, let $(\boldsymbol{\alpha}, \boldsymbol{\beta}, \boldsymbol{\gamma})^{\xi(e)} = (\boldsymbol{\alpha}', \boldsymbol{\beta}', \boldsymbol{\gamma}')$. Then*

$$Q(G; (\boldsymbol{\alpha}, \boldsymbol{\beta}, \boldsymbol{\gamma}), t) = \alpha'_e Q(G^{\xi(e)}/e; (\boldsymbol{\alpha}', \boldsymbol{\beta}', \boldsymbol{\gamma}'), t)$$
$$+ \beta'_e Q(G^{\xi(e)} - e; (\boldsymbol{\alpha}', \boldsymbol{\beta}', \boldsymbol{\gamma}'), t) + \gamma'_e Q(G^{\tau\xi(e)}/e; (\boldsymbol{\alpha}', \boldsymbol{\beta}', \boldsymbol{\gamma}'), t),$$

where on the right-hand side of the equation $(\boldsymbol{\alpha}', \boldsymbol{\beta}', \boldsymbol{\gamma}')$ is restricted to $G^{\xi(e)}/e$, $G^{\xi(e)} - e$, and $G^{\tau\xi(e)}/e$.

Proof. The identity follows easily from Theorems 4.10 and 4.12. $\qquad\square$

4.4 The Penrose Polynomial

The Penrose polynomial, $P(G, \lambda)$, of a plane graph first appeared (implicitly) in [84], where it arose out of Penrose's work on diagrammatic tensors. The Penrose polynomial has a number of remarkable graph theoretical properties, particularly with respect to graph colouring. For example, it is well known that proving the Four Colour Theorem is equivalent to showing that every plane, cubic, connected graph can be properly edge-coloured with three colours. The Penrose polynomial encodes exactly this information (see [84]): if G is a plane, cubic, connected graph, then

$$\text{the number of edge 3-colourings of } G = P(G; 3) = (-1/4)^{\frac{v(G)}{2}} P(G; -2). \quad (4.8)$$

The Penrose polynomial was recently extended to non-plane graphs in [28]. This extension led to a number of new properties of the Penrose polynomial of both plane graphs and embedded graphs. In this section we survey the Penrose polynomial and its extension to embedded graphs.

4.4.1 The Penrose Polynomial of an Embedded Graph and Its Relation to the Transition Polynomial

Referring to the notation from Sect. 1.5.2 if G_m is the canonically checkerboard coloured medial graph of G, then we let $\mathscr{P}(G_m)$ denote the set of states with no black smoothings, and we call such states *Penrose states*.

Definition 4.14. Let G be an embedded graph with canonically checkerboard coloured medial graph G_m, let $St(G_m)$ be the set of states of G_m, and let $\mathscr{P}(G_m)$ be the set of Penrose states. Then the *Penrose polynomial* is defined by

$$P(G;\lambda) := \sum_{s \in St(G_m)} \omega_P(s)\lambda^{c(s)} = \sum_{s \in \mathscr{P}(G_m)} (-1)^{cr(s)} \lambda^{c(s)} \in \mathbb{Z}[\lambda],$$

where $\omega_P(s)$ denotes the graph state weights of the medial weight system $W_P(G_m)$ defined by $(1, 0, -1)$, where $c(s)$ is the number of components in the graph state s, and where $cr(s)$ is the number of crossing vertex states in the graph state s.

Figure 4.5 gives an example of the computation of the Penrose polynomial.

The classical Penrose polynomial is just Definition 4.14 restricted to plane graphs. Since the topological Penrose polynomial is such a natural extension of the classical Penrose polynomial, we generally use the unmodified term Penrose polynomial for both and only use the adjectives when needed for emphasis.

The following proposition will be useful later. It expresses the Penrose polynomial as a sum over all of the partial Petrials, $G^{\tau(A)}$, of G.

Proposition 4.15. *If G is a ribbon graph, then*

$$P(G;\lambda) = \sum_{A \subseteq E(G)} (-1)^{|A|}\lambda^{f(G^{\tau(A)})},$$

where $f(G^{\tau(A)})$ is the number of faces of $G^{\tau(A)}$.

Proof. Let G_m be naturally embedded in the same surface as G. To each Penrose state s of G_m associate a set $A_s \subseteq E(G)$ by including an edge e in A_s if and only if the vertex state s_e at the vertex v_e of G_m corresponding to e is a crossing. This gives a bijection between the set of Penrose states and the set of subsets of $E(G)$. Clearly $cr(s) = |A_s|$. To prove the lemma we must show that $c(s) = f(G^{\tau(A_s)})$. To see why this is the case, let a, b, c, d be the "corners" of e. If s_e is a crossing that forms arcs (ac) and (bd) along e, then in $G^{\tau(A_s)}$, the boundary of the half-twisted edge e also forms arcs (ac) and (bd). Similarly, if s_e is a white smoothing that forms arcs (ad) and (bc) along e, then in $G^{\tau(A_s)}$, the boundary of the edge e also forms arcs (ad) and (bc). Since the points a, b, c, d for each edge e are connected along the vertex set of G in an identical way for both $G^{\tau(A_s)}$ and s, it follows that $c(s) = f(G^{\tau(A_s)})$. The proposition then follows. $\qquad\square$

The Penrose polynomial of an embedded graph can be written in terms of the topological transition polynomial. A comparison between the state sum relation for

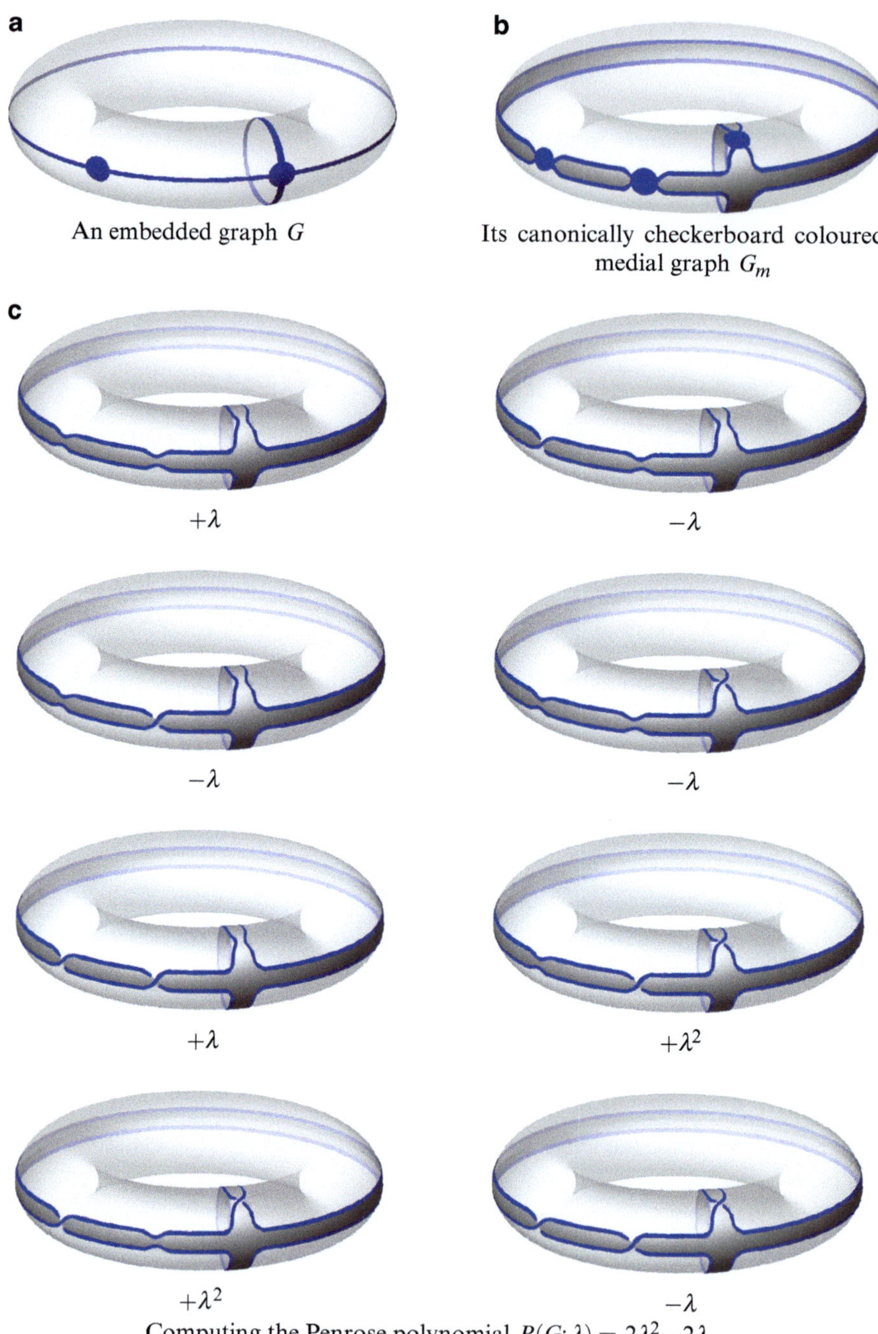

Computing the Penrose polynomial $P(G;\lambda) = 2\lambda^2 - 2\lambda$

Fig. 4.5 Computing the Penrose polynomial of an embedded graph

the Penrose polynomial in Definition 4.14, and that for the topological transition polynomial in Definitions 4.6 and 4.7, immediately makes clear how the Penrose polynomial is related to the topological transition polynomial:

Proposition 4.16. *Let G be an embedded graph and G_m be its canonically checkerboard coloured medial graph. Then*

$$P(G;\lambda) = q(G_m; W_P, \lambda) = Q(G; (1,0,-1), \lambda),$$

where Q is the topological transition polynomial. $\square$

A consequence of this is that since $Q(G)$ can be computed by linear recursion, the Penrose polynomial can also be computed via a linear recursion relation by repeatedly applying the skein relation

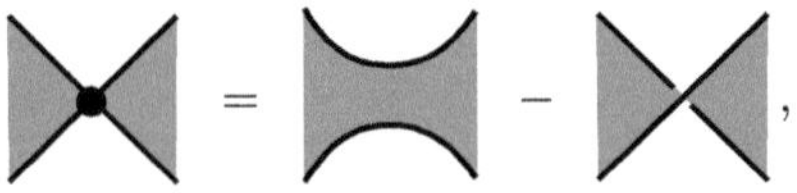

$$\tag{4.9}$$

to vertices of degree 4 in G_m, and at the end, evaluating each of the resulting closed curves to λ.

4.4.2   Identities for the Topological Penrose Polynomial

We will now see a number of identities for the Penrose polynomial that emerge from properties of twisted duality and the topological transition polynomial. This method of using the relation between the transition polynomial and the Penrose polynomial has antecedents in the plane case in [31].

By considering the broader class of embedded graphs, rather than just the class of plane graphs for which the Penrose polynomial was originally defined, we see that the Penrose polynomial has a recursion relation like that for the Tutte polynomial.

Theorem 4.17. *Suppose G is an embedded graph, and $e \in E(G)$. Then*

$$P(G;\lambda) = P(G/e;\lambda) - P(G^{\tau(e)}/e;\lambda).$$

Proof. The result follows immediately from Theorem 4.10, Corollary 4.13, and the definition of contraction. $\square$

Many graph polynomials satisfy natural duality relations. For example the Tutte polynomial and the topological transition polynomial do. We will now give a duality, or rather a twisted duality, identity for the Penrose polynomial.

Theorem 4.18. *Let G be an embedded graph and e be any edge of G. Then the Penrose polynomial satisfies the following twisted duality relation:*

$$P(G;\lambda) = P(G^{\delta(e)};\lambda) - P(G^{\delta\tau(e)};\lambda).$$

Proof. To show that $P(G;\lambda) = P(G^{\delta(e)};\lambda) - P(G^{\delta\tau(e)};\lambda)$, we begin by verifying the equation

$$q(G_m;W_P,\lambda) = q(G_m;W_P^{\delta(e)},\lambda) - q(G_m;W_P^{\tau\delta(e)},\lambda). \tag{4.10}$$

To see why this equation holds, note that by applying Proposition 4.8 to the weight systems W_P, $W_P^{\delta(e)}$, and $W_P^{\tau\delta(e)}$, respectively, we have

$$q(G_m;W_P,\lambda) = q((G_m)_{wh(v_e)};W_P,\lambda) - q((G_m)_{cr(v_e)};W_P,\lambda),$$

$$q(G_m;W_P^{\delta(e)},\lambda) = q((G_m)_{bl(v_e)};W_P,\lambda) - q((G_m)_{cr(v_e)};W_P,\lambda),$$

and

$$q(G_m;W_P^{\tau\delta(e)},\lambda) = -q((G_m)_{wh(v_e)};W_P,\lambda) + q((G_m)_{bl(v_e)};W_P,\lambda).$$

Substituting the three identities above into the left-hand and right-hand sides of Eq. (4.10) verifies the equality.

We will now express each of the terms in Eq. (4.10) in terms of the Penrose polynomial. By Proposition 4.16, we have

$$P(G;\lambda) = q(G_m;W_P,\lambda). \tag{4.11}$$

For the second term we have

$$\begin{aligned}
P(G^{\delta(e)};\lambda) &= q((G^{\delta(e)})_m;W_P,\lambda) \\
&= q(((G^{\delta(e)})^{\delta(e)})_m;W_P^{\delta(e)},\lambda) \\
&= q(G_m;W_P^{\delta(e)},\lambda),
\end{aligned} \tag{4.12}$$

where the first equality follows by Proposition 4.16, the second follows from Theorem 4.12, and the third follows from the fact that $(\delta(e))(\delta(e)) = 1(e)$.

Finally, we can rewrite the third term of Eq. (4.10) as follows:

$$\begin{aligned}
P(G^{\delta\tau(e)};\lambda) &= q((G^{\delta\tau(e)})_m;W_P,\lambda) \\
&= q(((G^{\delta\tau(e)})^{\tau\delta(e)})_m;W_P^{\tau\delta(e)},\lambda) \\
&= q(G_m;W_P^{\tau\delta(e)},\lambda).
\end{aligned} \tag{4.13}$$

Here, the first equality follows by Proposition 4.16, the second follows from Theorem 4.12, and the third follows from the fact that $(\delta\tau(e))(\tau\delta(e)) = 1(e)$.

The theorem then follows by substituting the identities in Eqs. (4.11)–(4.13) into Eq. (4.10). $\qquad\square$

The Penrose polynomial is also well behaved under partial Petrailty.

Proposition 4.19. *Let G be an embedded graph and $A \subseteq E(G)$. Then $P(G;\lambda) = (-1)^{|A|} P(G^{\tau(A)};\lambda)$ and in particular $|P(G;\lambda)|$ is invariant under partial Petriality.*

Proof. We have that

$$P(G^{\tau(A)};\lambda) = Q(G^{\tau(A)};(\mathbf{1},\mathbf{0},-\mathbf{1}),\lambda)$$
$$= Q((G^{\tau(A)})^{\tau(A)};(\mathbf{1},\mathbf{0},-\mathbf{1})^{\tau(A)},\lambda)$$
$$= Q(G;(\mathbf{1},\mathbf{0},-\mathbf{1})^{\tau(A)},\lambda),$$

where the first equality follows from Proposition 4.16, the second from Theorem 4.12, and the third from the fact that $\tau(A)\tau(A) = 1(A)$. It remains to show that

$$P(G^{\tau(A)};\lambda) = (-1)^{|A|} Q(G;(\mathbf{1},\mathbf{0},-\mathbf{1}),\lambda).$$

This identity follows by recalling the skein definition of the Penrose weight system to get

$$W_P^{\tau(A)}(G_m): \quad \bowtie_{v_e} \; = \; \begin{cases} \;\asymp \;-\; \bowtie & \text{if } e \notin A \\[2ex] \;\bowtie\; -\; \asymp & \text{if } e \in A \end{cases}$$

and observing that the relations differ only by a factor of -1. $\qquad\square$

Proposition 4.19 gives a quick proof of the following fact.

Corollary 4.20. *If G is a self-Petrial embedded graph, then $P(G;\lambda) = 0$ or $E(G)$ is even.*

Proof. G is self-Petrial if and only if $G^{\tau(E(G))} = G$. By Proposition 4.19, we then have $P(G;\lambda) = (-1)^{|E(G)|} P(G^{\tau(E(G))};\lambda)$, and the result follows. $\qquad\square$

Corollary 4.21. *If G is a self-Petrial plane graph with an odd number of edges, then G contains a bridge.*

Proof. It can be shown (see [2]) that when G be a plane graph, $P(G;\lambda) = 0$ if and only if G has a bridge. But if G is plane, bridgeless, self-Petrial with an odd number of edges, then since $P(G;\lambda) = (-1)^{|E(G)|} P(G^{\tau(E(G))};\lambda)$, we have $P(G;\lambda) = 0$, a contradiction. $\qquad\square$

Example 4.22. In the proof of Corollary 4.21, it was noted that a plane graph has trivial Penrose polynomial if and only if it contains a bridge. While $P(G;\lambda) = 0$ for all embedded graphs that contain a bridge, the converse does not hold in general. That is, a trivial Penrose polynomial does not ensure that an embedded graph contains a bridge. For example, the hemidodecahedron is bridgeless, but, as it has an odd number of edges and is self-Petrial, it follows from Corollary 4.20 that its Penrose polynomial is zero.

4.4.3 k-Valuations and the Penrose Polynomial

We will now give combinatorial interpretations for the Penrose polynomial evaluated at integer values. These interpretations are in terms of special edge colourings of the medial graph.

Definition 4.23. Let G be an embedded graph and G_m be its canonically checkerboard coloured medial graph. A *k-valuation* of G_m is a edge k-colouring $\phi : E(G_m) \to \{1,2,\ldots,k\}$ such that for each i and every vertex v_e of G_m, the number of i-coloured edges incident with v_e is even.

A k-valuation is said to be *admissible* if, at each vertex of G_m, the k-valuation is of one of the following two types:

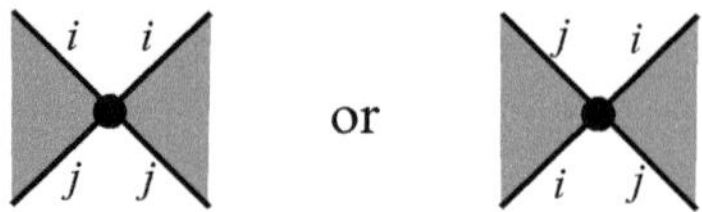

where $i \neq j$. The configuration on the left corresponds to a white smoothing state and the one on the right to a crossing state.

A k-valuation is said to be *permissible* if at each vertex of G_m, the k-valuation is of one of the two types shown above, or of the following type, called *total*:

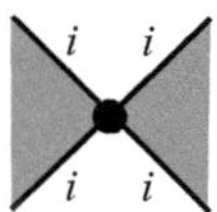

The following theorem shows that the Penrose polynomial evaluated at a positive integer n is a signed count of admissible n-valuations.

Theorem 4.24. *If G is an embedded graph and $n \in \mathbb{N}$, then $P(G;n) = \sum (-1)^{cr(s)}$, where the sum is over all admissible n-valuations s of G_m and where $cr(s)$ is the number of crossing states in s.*

Proof. We argue by induction on the number of edges of G. The result clearly holds if G has no edges, so let e be an edge of G. We note that the admissible colourings of a graph are unaffected by adding half-twists to edges, since the cyclic order of edges about each vertex is preserved. Thus, recalling Proposition 4.5, the admissible colourings of $(G_m)_{cr(v_e)}$ and $(G^{\tau(e)}/e)_m$ are the same.

By Theorem 4.17,

$$P(G;\lambda) = P(G/e;\lambda) - P(G^{\tau(e)}/e;\lambda).$$

By induction, the right-hand side is

$$\sum_{(G/e)_m} (-1)^{cr(s)} - \sum_{(G^{\tau(e)}/e)_m} (-1)^{cr(s)},$$

where the first sum is over all admissible n-valuations s of $(G/e)_m$ and the second sum is over all admissible n-valuations s of $(G^{\tau(e)}/e)_m$.

As noted above, this becomes

$$\sum_{(G_m)_{wh(v_e)}} (-1)^{cr(s)} - \sum_{(G_m)_{cr(v_e)}} (-1)^{cr(s)},$$

where the first sum is over all admissible n-valuations s of $(G_m)_{wh(v_e)}$ and the second sum is over all admissible n-valuations s of $(G_m)_{cr(v_e)}$.

In an admissible n-valuation of $(G_m)_{wh(v_e)}$, the two arcs forming the white smoothing can either be assigned different colours or both be assigned the same colour and similarly in the crossing case. We can separate the sums accordingly:

$$\sum_{\substack{(G_m)_{wh(v_e)} \\ \text{diff}}} (-1)^{cr(s)} + \sum_{\substack{(G_m)_{wh(v_e)} \\ \text{same}}} (-1)^{cr(s)} - \sum_{\substack{(G_m)_{cr(v_e)} \\ \text{diff}}} (-1)^{cr(s)} - \sum_{\substack{(G_m)_{cr(v_e)} \\ \text{same}}} (-1)^{cr(s)}.$$

The first sum corresponds to a sum over all admissible n-valuations of G_m with a smoothing state at v_e. The second and fourth sums both correspond to all permissible n valuations of G_m with exactly one total vertex at v_e and hence cancel. The third sum corresponds to a sum over all admissible n-valuations of G_m with a crossing state at v_e, but with one more crossing to be counted, thus changing the subtraction to addition. This yields $P(G;n) = \sum (-1)^{cr(s)}$, as desired. $\qquad\square$

For plane graphs, Theorem 4.24 specialises to show that $P(G;n)$ counts the number of n-valuations of G_m.

Theorem 4.25 (Jaeger [50]). *If G is a plane graph, then for each $n \in \mathbb{N}$, $P(G;n)$ is equal to the number of admissible n-valuations of G_m.*

Proof. If G is plane, and E_i is the subset of edges of colour i in any admissible colouring, then $G_m|_{E_i}$ is a disjoint union of simple closed curves in the plane. By the Jordan Curve Theorem, the intersection of $G_m|_{E_i}$ and $G_m|_{E_j}$ for any $i \neq j$ must be an even number of points. Thus, $cr(s) \equiv 0 \bmod 2$ for every admissible colouring s of G_m. $\qquad\square$

We have found a combinatorial interpretation for the Penrose polynomial evaluated at positive integers. Below we state without proof a combinatorial interpretation for the Penrose polynomial evaluated at negative integers from [28]. It was proved by relating the Penrose polynomial to the circuit partition polynomial.

Theorem 4.26. *If G is orientable and checkerboard colourable, then $P(G;-n) = (-1)^{f(G)} \sum 2^{m(s)}$, where the sum is over all permissible n-valuations s of G_m and $m(s)$ is the number of total vertices in s.*

Theorem 4.26 does not hold for general embedded graphs. For example, any graph with a bridge has a Penrose polynomial of zero.

4.4.4 Graph Colouring and the Penrose Polynomial

In [2], Aigner proved the following result which connects the Penrose polynomial of a plane graph, G, and the chromatic polynomial of its geometric dual G^*.

Theorem 4.27 (Aigner [2]). *Let G be a plane graph, then for all $k \in \mathbb{N}$ we have*

$$\chi(G^*;k) \leq P(G;k). \tag{4.14}$$

In this section we will see that twisted duality enables us to complete Aigner's relation (4.14) between the Penrose and chromatic polynomials. In Theorem 4.28, we write the Penrose polynomial of an embedded graph as a signed sum of the chromatic polynomials of its twisted duals. We then find a simpler expression without the signs in the special case of plane graphs. $\chi(G^*;k)$ appears as a single summand in this expression, and so Theorem 4.27 follows from it.

Theorem 4.28. *Let G be an embedded graph. Then*

$$P(G;\lambda) = \sum_{A \subseteq E(G)} (-1)^{|A|} \chi((G^{\tau(A)})^*;\lambda).$$

Proof. By Theorem 4.24, for each $n \in \mathbb{N}$ we have

$$P(G;n) = \sum_{\substack{s \text{ an admiss.} \\ n\text{-val. of } G_m}} (-1)^{cr(s)} = \sum_{X \subseteq V(G_m)} (-1)^{|X|} \sum_{\substack{\text{admiss. } n\text{-val.} \\ \text{with crossing} \\ \text{states at } v \in X}} 1.$$

We now view G as a ribbon graph and construct a bijection from the set of admissible n-valuations of the medial graph G_m to a certain set of colourings of the boundaries of the partial Petrials of G. Let $A_s \subseteq E(G)$ be the set of edges corresponding to vertices of G_m with crossings in the state s. The cycles in G_m (which are determined by the colours in the n-valuation s) follow exactly the boundary components of the partial Petrial $G^{\tau(A_s)}$. Moreover, the colours of the cycles in the n-valuation induce a colouring of the boundary components of $G^{\tau(A_s)}$. We define a *proper boundary n-colouring* of a ribbon graph to be a map from its set of boundary components to the colours $\{1,2,\ldots,n\}$ with the property that whenever two boundary components share a common edge, they are assigned different colours. It is then clear that the map from s to $G^{\tau(A_s)}$ defines a bijection between the set of admissible n-valuations of G_m and the set of proper boundary n-colourings of the partial Petrials of G. Moreover, this bijection identifies the vertices with crossing states with the edges in A_s. Thus we have

$$\sum_{X \subseteq V(G_m)} (-1)^{|X|} \sum_{\substack{\text{admiss. } n\text{-val.} \\ \text{with crossing} \\ \text{states at } v \in X}} 1 = \sum_{A \subseteq E(G)} (-1)^{|A|} \sum_{\substack{\text{proper boundary} \\ n\text{-col. of } G^{\tau(A)}}} 1. \tag{4.15}$$

A proper boundary n-colouring of $G^{\tau(A)}$ corresponds to a proper face n-colouring of $G^{\tau(A)}$, and hence to a proper n-colouring of $(G^{\tau(A)})^*$. Since the chromatic polynomial counts proper n-colourings of a graph, we can then write Eq. (4.15) as

$$\sum_{A \subseteq E(G)} (-1)^{|A|} \sum_{\substack{\text{proper face} \\ n\text{-col. of } G^{\tau(A)}}} 1 = \sum_{A \subseteq E(G)} (-1)^{|A|} \sum_{\substack{\text{proper } n\text{-col.} \\ \text{of } (G^{\tau(A)})^*}} 1$$

$$= \sum_{A \subseteq E(G)} (-1)^{|A|} \chi((G^{\tau(A)})^*; n).$$

The theorem then follows upon observing that n is an arbitrary natural number. $\square$

Theorem 4.28 can be simplified when G is a plane graph. To do this, we will need the following lemma.

Lemma 4.29. *If G is a plane graph and $(G^{\tau(A)})^*$ is loopless, then A is even.*

Proof. Embed G (as a ribbon graph) on the plane. Using the same drawing, add half-twists to all of the edges in A, to get a drawing (but not an embedding) of $G^{\tau(A)}$ on the plane. Note that the set S of boundary components of the drawing of $G^{\tau(A)}$ is a set of closed plane curves. If $(G^{\tau(A)})^*$ does not contain a loop, then all of the curves in S are simple. By the Jordan Curve Theorem, if $c, c' \in S$, then they intersect an even number of times (every time c enters c', it has to exit it). So the total number of intersection points between pairs of curves in S is even. But, as the curves in S are simple, these account for all of the intersection points. Finally, the intersection points are in correspondence with edges in A, so A is even. $\square$

Corollary 4.30. *Let G be a plane graph. Then*

$$P(G; \lambda) = \sum_{A \subseteq E(G)} \chi((G^{\tau(A)})^*; \lambda). \tag{4.16}$$

Proof. By Lemma 4.29, if $|A|$ is odd, then $(G^{\tau(A)})^*$ must contain a loop. Thus, $(-1)^{|A|} = 1$ or $\chi((G^{\tau(A)})^*; n) = 0$. The result then follows from Theorem 4.28. $\square$

Note that Theorem 4.27 follows immediately from Corollary 4.30 as $\chi((G^{\tau(A)})^*)$ appears as a single summand in the right-hand side of Eq. (4.16).

We conclude this section by using Corollary 4.30 to reformulate the Four Colour Theorem.

Corollary 4.31. *The following statements are equivalent:*

1. *The Four Colour Theorem is true.*
2. *For every connected, bridgeless plane graph G there exists $A \subseteq E(G)$ such that $\chi((G^{\tau(A)})^*; 3) \neq 0$.*
3. *For every connected, bridgeless plane graph G there exists $A \subseteq E(G)$ such that $\chi((G^{\tau(A)})^*; 4) \neq 0$.*

Proof. Corollary 9 of [2] states that proving the Four Colour Theorem is equivalent to showing that $P(G;3) > 0$ or $P(G;4) > 0$ for all connected, bridgeless plane graphs G. Since $\chi(H;k) \geq 0$ for all $k \in \mathbb{N}$ and graphs H, Corollary 4.30 tells us that $P(G;k) > 0$ if and only if one of the summands $\chi((G^{\tau(A)})^*;k) \neq 0$. The result then follows. $\square$

4.5 Topological Tutte Polynomials

There have been a number of extensions of the Tutte polynomial to graphs embedded in surfaces, which we refer loosely to as *topological Tutte polynomials*. The first such extension appeared in Las Vergnas' work in 1978 (see [65–67]). However, most of the recent interest in this area was spurred by a (different) topological Tutte polynomial, Bollobás and Riordan's ribbon graph polynomial, $R(G)$, of [5, 6]. Much attention has been given to understanding how properties of the Tutte polynomial $T(G)$ extend to its topological analogue $R(G)$. Properties of the Tutte polynomial that are known to extend include duality relations [34,75], a quasi-tree expansion [15,96], Brylawski's tensor product formula [43], the recipe theorem [34], and connections with knot theory [8,17,18,21,75]. However, remarkably little is yet known about what information is encoded by topological Tutte polynomials.

In the remainder of this chapter we study the ribbon graph polynomial and its extension to edge weighted embedded graphs. In particular, we relate $R(G)$ to the topological transition polynomial and use this relation, in this and the next section, to find duality relations, relations with the Penrose polynomial, and to reveal some of its combinatorial and topological properties. In the next chapter we will consider its connections with knot polynomials.

4.5.1 *The Ribbon Graph Polynomial and the Topochromatic Polynomial*

The ribbon graph polynomial of Bollobás and Riordan, $R(G)$, extends the Tutte polynomial by adding terms to the state sum expansion of the Tutte polynomial (Eq. (4.3)) that encode some topological information (specifically the Euler genus and orientability) about the spanning embedded subgraphs. Then the topochromatic polynomial $Z(G)$ incorporates edge weights into the formulation. As with the Tutte polynomial, the ribbon graph polynomial satisfies some deletion–contraction relations. However, these relations do not apply to all edges of the embedded graph. In this section we discuss deletion–contraction relations for these polynomials, give relations for the topological transition polynomial, and describe the duality and partial duality properties of the polynomials.

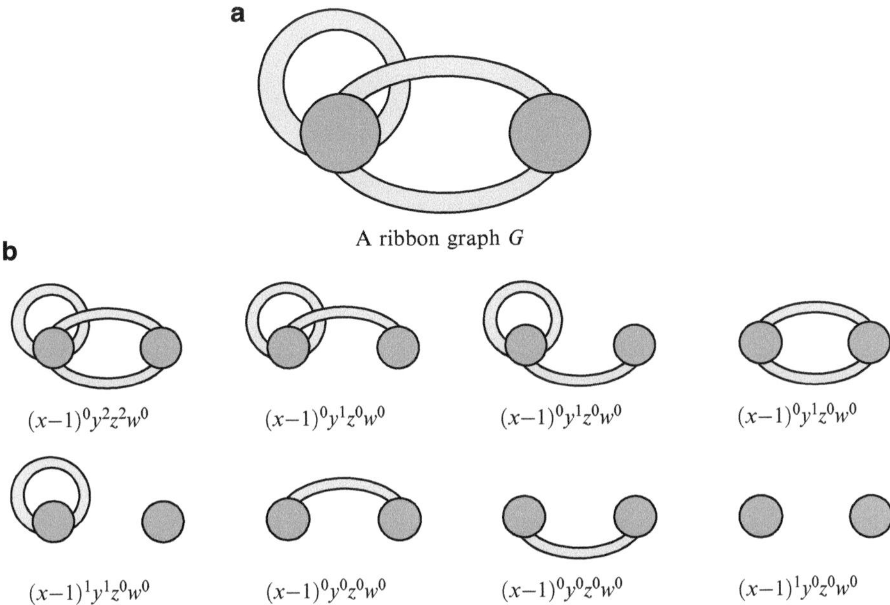

Fig. 4.6 Computing the ribbon graph polynomial $R(G;x,y,z,w)$ of an embedded graph

As mentioned above, the ribbon graph polynomial $R(G)$ is defined by a state sum expansion. For this state sum expansion, in addition to the graph parameters described in Chap. 1, we need $t(G)$, which records the orientability of an embedded graph G by $t(G) = 0$ if G is orientable and $t(G) = 1$ otherwise.

Definition 4.32 (Bollobás and Riordan [6]). Let G be an embedded graph. Then the *ribbon graph polynomial*, $R(G;x,y,z,w) \in \mathbb{Z}[x,y,z,w]/\langle w^2 - w \rangle$, is defined by

$$R(G;x,y,z,w) = \sum_{A \subseteq E(G)} (x-1)^{r(G)-r(A)} y^{n(A)} z^{k(A)-f(A)+n(A)} w^{t(A)}.$$

The ribbon graph polynomial is also known in the literature as the *Bollobás–Riordan polynomial*. An example of computing $R(G)$ is given in Fig. 4.6.

The polynomial $R(G)$ contains the (classical) Tutte polynomial $T(G)$ as a specialisation and agrees with it on plane graphs:

Proposition 4.33 (Bollobás and Riordan [6]). *Let G be an embedded graph. Then*

1. $T(G;x,y) = R(G;x,y-1,1,1)$.
2. If G is plane, then $T(G;x,y) = R(G;x,y-1,z,w)$ for all z and w.

Proof. The first item follows immediately upon comparing the state sum for $R(G;x,y-1,1,1)$ with that of $T(G;x,y)$ from Eq. (4.3).

For the second item, observe that

$$k(A) - f(A) + n(A) = 2k(A) - v(A) + e(A) - f(A) = \gamma(A),$$

where γ is the Euler genus. Therefore, if G is plane, $z^{k(A)-f(A)+n(A)} = z^0$ and $w^{t(A)} = w^0$ for all $A \subseteq E(G)$, and the result follows. $\square$

The polynomial $R(G)$ has been extended by the topochromatic polynomial, a multivariate generalisation in [75]. We will usually work with the topochromatic polynomial Z here, translating results back to R as needed using Proposition 4.37 below.

Definition 4.34 ([75]). Let G be an embedded graph. Let a, c, and w be indeterminates, and let $b := \{b_e | e \in E(G)\}$ be a set of indeterminates indexed by $E(G)$. The *topochromatic polynomial* is

$$Z(G;a,b,c,w) = \sum_{A \subseteq E(G)} a^{k(A)} \left(\prod_{e \in A} b_e \right) c^{f(A)} w^{t(A)} \in \mathbb{Z}[a,b,c,w]/\langle w^2 - w \rangle.$$

Notation 4.35. If all of the edge weights in b are equal, say $b_e = k$ for all e, then we will often write $Z(G;a,k,c,w)$ for $Z(G;a,b,c,w)$.

$Z(G)$ is often known in the literature as the multivariate Bollobás–Riordan polynomial to emphasise its connection to the Bollobás–Riordan ribbon graph polynomial. However the name topochromatic polynomial used here as it recalls that $Z(G)$ is a multivariate extension to embedded graphs of the dichromatic polynomial or random cluster model, $Z_T(G;u,v) = \sum u^{k(A)} v^{|A|}$.

Example 4.36. If G is the ribbon graph shown in Fig. 2.11a, and $b = \{b_1, b_2, b_3\}$, then

$$Z(G;a,b,c,w) = ab_1 b_2 b_3 cw + ab_1 b_2 cw + ab_1 b_3 cw + ab_2 b_3 c^2$$

$$+ a^2 b_1 c^2 w + ab_2 c + ab_3 c + a^2 c^2.$$

The following result can be obtained by expanding the rank and nullity terms in the definition of $R(G)$.

Proposition 4.37. *The ribbon graph and topochromatic polynomials are related as follows:*

$$R(G;x,y,z,w) = (x-1)^{-k(G)}(yz)^{-v(G)} Z(G;(x-1)yz^2, b, z^{-1}, w), \qquad (4.17)$$

where the edge weights b are given by setting $b_e = yz$ for each $e \in E(G)$. Also,

$$Z(G;a,b,c,w) = (ac/b)^{k(G)} b^{v(G)} R(G;(ac+b)/b, bc, 1/c, w), \qquad (4.18)$$

where the edge weights b are given by setting $b_e = b$ for each $e \in E(G)$.

The following theorem provides deletion–contraction relations that act on all of the edges of an embedded graph except for non-trivial orientable loops. This means that the relations will reduce any embedded graph to a linear combination of disjoint unions of orientable one-vertex embedded graphs in which every edge alternates with another. However, the relations do not provide a way to calculate Z of such embedded graphs, so the theorem does not provide a full recursive definition of $Z(G)$. At the end of the recursion, Definition 4.34 must be applied to compute $Z(G)$. For example, if G has one vertex and two loops and is embedded on a torus, then none of the relations in Theorem 4.38 apply, so Definition 4.34 must be used to find that $Z(G) = ab^2c + 2abc^2 + ac$. An example of using the deletion–contraction relations of Theorem 4.38 to compute $Z(G)$ is given in Fig. 4.7. This should be compared with Example 4.36. We remind the reader that $G/e := G^{\delta(e)} - e$.

Theorem 4.38. *Let G and H be embedded graphs, and $e \in E(G)$. Then*

1. If e is not a loop,

$$Z(G;a,\boldsymbol{b},c,w) = Z(G - e;a,\boldsymbol{b},c,w) + b_e Z(G/e;a,\boldsymbol{b},c,w)$$

2. If e is a trivial orientable loop,

$$Z(G;a,\boldsymbol{b},c,w) = Z(G - e;a,\boldsymbol{b},c,w) + a^{-1} b_e Z(G/e;a,\boldsymbol{b},c,w)$$

3. If e is a non-orientable loop, trivial or not,

$$Z(G;a,\boldsymbol{b},c,w) = Z(G - e;a,\boldsymbol{b},c,w) + b_e w Z(G/e;a,\boldsymbol{b},c,w);$$

4. $Z(G \sqcup H) = Z(G)Z(H) = (ac)Z(G \vee H)$

Proof. We first prove the deletion–contraction properties.

By partitioning the subsets of $E(G)$ according to those that contain e and those that do not, we can write

$$Z(G;a,\boldsymbol{b},c,w) = \sum_{\substack{A \subseteq E(G) \\ e \notin A}} a^{k(A)} \left(\prod_{e \in A} b_e \right) c^{f(A)} w^{t(A)} + \sum_{\substack{A \subseteq E(G) \\ e \in A}} a^{k(A)} \left(\prod_{e \in A} b_e \right) c^{f(A)} w^{t(A)}.$$

$$(4.19)$$

We will show that the first sum on the right-hand side of Eq. (4.19) is equal to $Z(G - e;a,\boldsymbol{b},c,w)$ and will relate the second sum to $Z(G/e;a,\boldsymbol{b},c,w)$.

Define a mapping φ from the set of spanning subgraphs of G that do not contain e and the set of spanning subgraphs of $G - e$ by $\varphi : (V(G),A) \mapsto (V(G - e),A)$. Clearly, φ is a bijection. Also, if $H = (V(G),A)$, with $e \notin A$, then the following hold:

$$k(H) = k(\varphi(H)), \qquad\qquad f(H) = f(\varphi(H)),$$

$$w(H) = w(\varphi(H)), \qquad\qquad \prod_{f \in E(H)} b_f = \prod_{f \in E(\varphi(H))} b_f.$$

Thus, for each $H = (V(G), A)$, with $e \notin A$,

$$a^{k(H)}\left(\prod_{f \in E(H)} b_f\right) c^{f(H)} w^{t(H)} = a^{k(\varphi(H))}\left(\prod_{f \in E(\varphi(H))} b_f\right) c^{f(\varphi(H))} w^{t(\varphi(H))},$$

and so

$$\sum_{\substack{A \subseteq E(G) \\ e \notin A}} a^{k(A)}\left(\prod_{f \in A} b_f\right) c^{f(A)} w^{t(A)} = \sum_{A \subseteq E(G-e)} a^{k(A)}\left(\prod_{f \in A} b_f\right) c^{f(A)} w^{t(A)} \quad (4.20)$$

$$= Z(G - e; a, \boldsymbol{b}, c, w).$$

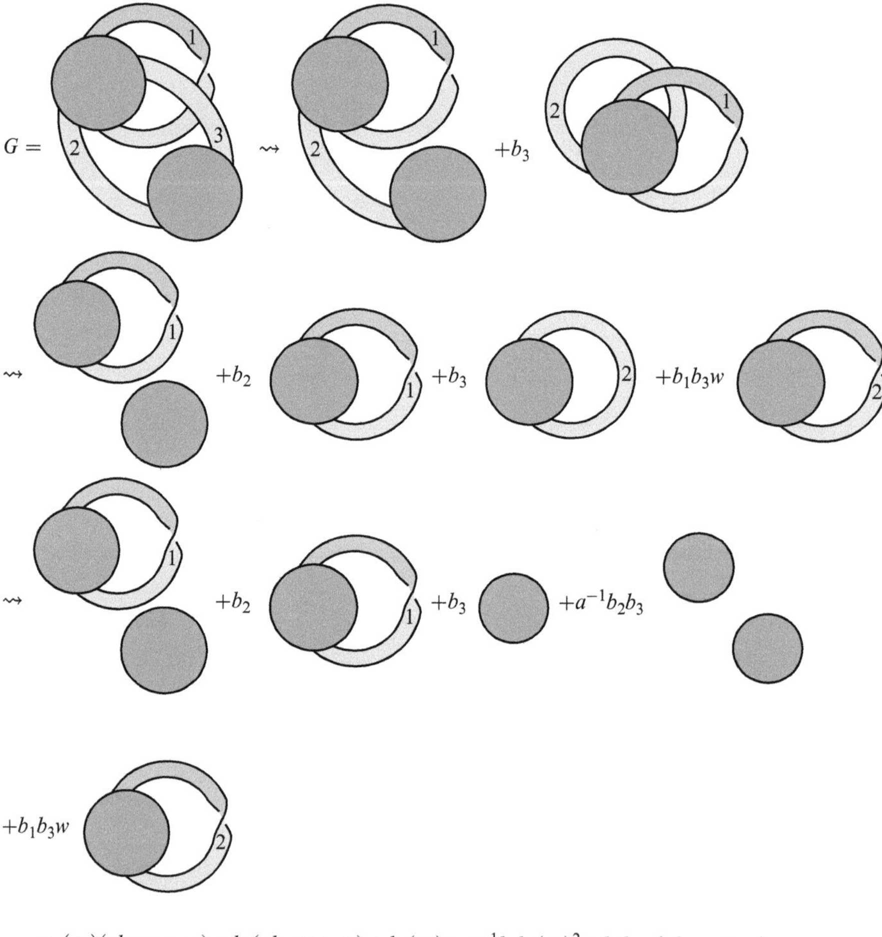

$$\rightsquigarrow (ac)(ab_1 cw + ac) + b_2(ab_1 cw + ac) + b_3(ac) + a^{-1}b_2 b_3(ac)^2 + b_1 b_3 w(ab_2 cw + ac)$$
$$= ab_1 b_2 b_3 cw + ab_1 b_2 cw + ab_1 b_3 cw + ab_2 b_3 c^2 + a^2 b_1 c^2 w + ab_2 c + ab_3 c + a^2 c^2 = Z(G; a, \boldsymbol{b}, c, w)$$

Fig. 4.7 Computing $Z(G)$ using its deletion–contraction relations

Now define a mapping ψ from the set of spanning subgraphs of G that contain e and the set of spanning subgraphs of G/e by $\psi : (V(G),A) \mapsto (V(G/e),A\backslash\{e\})$. Clearly, ψ is a bijection. Now suppose $H = (V(G),A)$ is a spanning subgraph of G with $e \in A$. Then $\psi(H) = H/e$. By examining Fig. 4.1 we see that the following properties hold:

$$f(H) = f(H/e) \quad \text{and} \quad \prod_{f \in E(H)} b_f = b_e \left(\prod_{f \in E(H/e)} b_f \right).$$

If e is not an orientable loop, then $k(H) = k(H/e)$. If e is an orientable trivial loop, then $k(H) = k(H/e) - 1$. If e is not a loop, or is a trivial orientable loop, then $t(H) = t(H/e)$. If e is a non-orientable loop, then H is non-orientable, and, since $w^2 = w$, we have $w^{t(H)} = w \cdot w^{t(H)} = w \cdot w^{t(H/e)}$. Observe that if e is a non-trivial orientable loop, then we cannot determine the exact relation between $k(H)$ and $k(H/e)$ or between $w(H)$ and $w(H/e)$.

Thus, for each $H = (V(G),A)$, with $e \in A$ and e not a loop

$$a^{k(H)} \left(\prod_{f \in E(H)} b_f \right) c^{f(H)} w^{t(H)} = b_e a^{k(H/e)} \left(\prod_{f \in E(H/e)} b_f \right) c^{f(H/e)} w^{t(H/e)},$$

so

$$\sum_{\substack{A \subseteq E(G) \\ e \notin A}} a^{k(A)} \left(\prod_{f \in A} b_f \right) c^{f(A)} w^{t(A)} = b_e \sum_{A \subseteq E(G/e)} a^{k(A)} \left(\prod_{f \in A} b_f \right) c^{f(A)} w^{t(A)}$$

$$= b_e Z(G/e;a,\boldsymbol{b},c,w). \tag{4.21}$$

For each $H = (V(G),A)$, with $e \in A$ and e a trivial orientable loop

$$a^{k(H)} \left(\prod_{f \in E(H)} b_f \right) c^{f(H)} w^{t(H)} = a^{-1} b_e a^{k(H/e)} \left(\prod_{f \in E(H/e)} b_f \right) c^{f(H/e)} w^{t(H/e)},$$

so

$$\sum_{\substack{A \subseteq E(G) \\ e \notin A}} a^{k(A)} \left(\prod_{f \in A} b_f \right) c^{f(A)} w^{t(A)} = a^{-1} b_e \sum_{A \subseteq E(G/e)} a^{k(A)} \left(\prod_{f \in A} b_f \right) c^{f(A)} w^{t(A)}$$

$$= a^{-1} b_e Z(G/e;a,\boldsymbol{b},c,w). \tag{4.22}$$

For each $H = (V(G),A)$, with $e \in A$ and e a non-orientable loop

$$a^{k(H)} \left(\prod_{f \in E(H)} b_f \right) c^{f(H)} w^{t(H)} = b_e w a^{k(H/e)} \left(\prod_{f \in E(H/e)} b_f \right) c^{f(H/e)} w^{t(H/e)},$$

so

$$\sum_{\substack{A \subseteq E(G) \\ e \notin A}} a^{k(A)} \left(\prod_{f \in A} b_f \right) c^{f(A)} w^{t(A)} = b_e w \sum_{A \subseteq E(G/e)} a^{k(A)} \left(\prod_{f \in A} b_f \right) c^{f(A)} w^{t(A)}$$

$$= b_e w Z(G/e; a, \boldsymbol{b}, c, w). \tag{4.23}$$

The results for Items 1 through 3 then follow by substituting Eqs. (4.20)–(4.23) into Eq. (4.19).

For Item 4,

$$Z(G \sqcup H; a, \boldsymbol{b}, c, w) = \sum_{A \subseteq E(G \sqcup H)} a^{k(A)} \left(\prod_{e \in A} b_e \right) c^{f(A)} w^{t(A)}$$

$$= \sum_{\substack{A \subseteq E(G) \\ B \subseteq E(H)}} a^{k(A \sqcup B)} \left(\prod_{e \in A \sqcup B} b_e \right) c^{f(S \sqcup B)} w^{t(A \sqcup B)}$$

$$= \sum_{A \subseteq E(G)} \sum_{B \subseteq E(H)} a^{k((V(G),A))} a^{k((V(H),B))} \left(\prod_{e \in A} b_e \right) \left(\prod_{e \in B} b_e \right)$$

$$\cdot c^{f((V(G),A))} c^{f((V(H),B))} w^{t((V(G),A))} w^{t((V(H),B))}$$

$$= \left(\sum_{A \subseteq E(G)} a^{k((V(G),A))} \left(\prod_{e \in A} b_e \right) c^{f((V(G),A))} w^{t((V(G),A))} \right)$$

$$\cdot \left(\sum_{B \subseteq E(H)} a^{k((V(H),B))} \left(\prod_{e \in B} b_e \right) c^{f((V(H),B))} w^{t((V(H),B))} \right)$$

$$= Z(G; a, \boldsymbol{b}, c, w) \cdot Z(H; a, \boldsymbol{b}, c, w).$$

The proof that $Z(G \vee H; a, \boldsymbol{b}, c, w) = (ac)^{-1} Z(G; a, \boldsymbol{b}, c, w) \cdot Z(H; a, \boldsymbol{b}, c, w)$ is similar and is left as an exercise. $\square$

The following theorem is due to Bollobás and Riordan [6], except in the case where e is a non-orientable non-trivial loop. Here it follows as a corollary of Theorem 4.38 and Proposition 4.37. It provides some additional properties of the ribbon graph polynomial, including deletion–contraction relations for certain edges of an embedded graph. Again, the identities in Items 1–4 in Theorem 4.39 will reduce an embedded graph to a $\mathbb{Z}[x, y, z, w]/\langle w^2 - w \rangle$-linear combination of R computed on disjoint unions of 1-vertex embedded graphs that contain only non-trivial orientable loops. The values of R on these disjoint unions 1-vertex graphs are found by using the state sum from Definition 4.32.

Theorem 4.39. *Let G and H be embedded graphs. Then*

1. If e is neither a loop nor a bridge,

$$R(G; x, y, z, w) = R(G - e; x, y, z, w) + R(G/e; x, y, z, w)$$

2. *If e is a bridge,*

$$R(G;x,y,z,w) = xR(G/e;x,y,z,w)$$

3. *If e is a trivial orientable loop*

$$R(G;x,y,z,w) = (1+y)R(G-e;x,y,z,w)$$

4. *If e is a non-orientable loop*

$$R(G;x,y,z,w) = R(G-e;x,y,z,w) + yzwR(G/e;x,y,z,w)$$

5. $R(G \sqcup H) = R(G)R(H) = R(G \vee H)$

Proof. By Eq. (4.17)

$$R(G;x,y,z,w) = (x-1)^{-k(G)}(yz)^{-v(G)}Z(G;(x-1)yz^2,yz,z^{-1},w).$$

We first prove Item 5, which uses Item 4 of Theorem 4.38.

$$R(G \sqcup H;x,y,z,w) = (x-1)^{-k(G \sqcup H)}(yz)^{-v(G \sqcup H)}Z(G \sqcup H;(x-1)yz^2,yz,z^{-1},w)$$

$$= (x-1)^{-k(G)}(yz)^{-v(G)}Z(G;(x-1)yz^2,yz,z^{-1},w)$$

$$\cdot (x-1)^{-k(H)}(yz)^{-v(H)}Z(H;(x-1)yz^2,yz,z^{-1},w)$$

$$= R(G;x,y,z,w) \cdot R(H;x,y,z,w).$$

The argument that $R(G \vee H;x,y,z,w) = R(G;x,y,z,w) \cdot R(H;x,y,z,w)$ follows similarly and is left as an exercise.

For the deletion–contraction relations, if e is not a non-trivial orientable loop, then Theorem 4.38 gives

$$(x-1)^{k(G)}(yz)^{v(G)}R(G;x,y,z,w)$$

$$= (x-1)^{k(G-e)}(yz)^{v(G-e)}R(G-e;x,y,z,w)$$

$$+ \alpha_e(yz)(x-1)^{k(G/e)}(yz)^{v(G/e)}R(G/e;x,y,z,w),$$

where

$$\alpha_e = \begin{cases} 1 & \text{if } e \text{ is a not a loop,} \\ ((x-1)yz^2)^{-1} & \text{if } e \text{ is a trivial orientable loop,} \\ w & \text{if } e \text{ is a non-orientable loop.} \end{cases}$$

Table 4.2 Counting vertices and components for the proof of Corollary 4.42

Type of edge	$k(G-e)-k(G)$	$v(G-e)-v(G)$	$k(G/e)-k(G)$	$v(G/e)-v(G)+1$
Non-loop, non-bridge	0	0	0	0
Bridge	1	0	0	0
Trivial orientable loop	0	0	1	2
Non-trivial orientable loop	0	0	0	2
Non-orientable loop	0	0	0	1

This gives

$$
R(G;x,y,z,w) = (x-1)^{k(G-e)-k(G)}(yz)^{v(G-e)-v(G)}R(G-e;x,y,z,w)
$$
$$
+\alpha_e(x-1)^{k(G/e)-k(G)}(yz)^{v(G/e)-v(G)+1}R(G/e;x,y,z,w). \quad (4.24)
$$

Table 4.2 shows the values of the normalisation factors for the various types of edge e. The theorem follows by substituting these values into Eq. (4.24) and by using the fact that $R(G-e) = R(G/e)$ when e is a bridge, or trivial orientable loop. $\qquad\square$

4.5.2 Relation to the Topological Transition Polynomial

Having reviewed $R(G)$ and $Z(G)$, we now establish their relation to the topological transition polynomial. To see this relation, we let G be an embedded graph and G_m be its embedded medial graph equipped with the canonical checkerboard colouring. We define the weight system $W_Z(G_m)$ by

$$
W_Z(G_m) : \quad \bowtie_{v_e} = b_e \;)(\; +1 \;)(\; +0 \; \bowtie
$$

Also, in analogy with Penrose states, we define the *topochromatic states* to be the graph states of G_m with no crossings and denote them by $\mathscr{Z}(G_m)$.

With this weight system we obtain a relation between $Z(G)$ and $Q(G)$.

Proposition 4.40. *If G is an embedded graph and $(\boldsymbol{b},\mathbf{1},\mathbf{0})$ corresponds to the weight system W_Z, then*

$$
Z(G;1,\boldsymbol{b},c,1) = q(G_m;W_Z,c) = Q(G;(\boldsymbol{b},\mathbf{1},\mathbf{0}),c).
$$

Proof. By definition,

$$
Q(G;(\boldsymbol{b},\mathbf{1},\mathbf{0}),c) = \sum_{s\in\mathscr{Z}(G_m)} \omega_Z(s)c^{c(s)} = \sum_{s\in\mathscr{Z}(G_m)} \left(\prod_{v_e\in Wh(s)} b_e \right)c^{c(s)},
$$

where $c(s)$ is the number of components in the state s, where $\omega_z(s)$ is the weight of the state s in the weight system W_Z, and where $Wh(s)$ is the set of vertices with white smoothing states in the graph state s.

We can define a bijection between the set of embedded spanning subgraphs of G and the set of graph states of G_m by associating an edge set $A_s \subseteq E(G)$ with a graph state s of G_m by setting $e \in A_s$ if and only if the vertex state $v_e \in Wh(s)$. It is then clear that, for every graph state, $c(s) = f(A_s)$. By using this bijection, we have

$$Q(G;(\boldsymbol{b},1,0),c) = \sum_{s \in \mathscr{L}(G_m)} \left(\prod_{v_e \in Wh(s)} b_e \right) c^{c(s)} = \sum_{A \subseteq E(G)} \left(\prod_{e \in A} b_e \right) c^{f(H)}$$

$$= Z(G;1,\boldsymbol{b},c,1). \qquad \square$$

The specialisation of Z to $Z(G;1,\boldsymbol{b},c,1)$, together with the properties of the topological transition polynomial, leads to a deletion–contraction relation, in Theorem 4.41, which, unlike the general case of Theorem 4.39, works for all edges, including non-trivial orientable loops. The relations of Theorem 4.41 reduce any ribbon graph to a linear combination of embedded graphs that consist only of vertices. As the theorem provides the values of the polynomial on these edgeless embedded graphs, it therefore provides a full recursive definition of $Z(G;1,\boldsymbol{b},c,1)$.

Theorem 4.41. *Let G be an embedded graph and e be any edge of G. Then*

$$Z(G;1,\boldsymbol{b},c,1) = Z(G-e;1,\boldsymbol{b},c,1) + b_e Z(G/e;1,\boldsymbol{b},c,1).$$

In addition, $Z(G \sqcup H) = Z(G)Z(H)$, and $Z(E;1,\boldsymbol{b},c,1) = c$, when $E = (\{v\},\emptyset)$.

Proof. By Proposition 4.40 and Theorem 4.10, we have

$$Z(G;1,\boldsymbol{b},c,1) = Q(G;(\boldsymbol{b},1,0),c)$$
$$= b_e Q(G/e;(\boldsymbol{b},1,0),c) + 1 \cdot Q(G-e;(\boldsymbol{b},1,0),c)$$
$$+ 0 \cdot Q(G^{\tau(e)}/e;(\boldsymbol{b},1,0),c)$$
$$= Z(G-e;1,\boldsymbol{b},c,1) + b_e Z(G/e;1,\boldsymbol{b},c,1).$$

Finally, that $Z(G \sqcup H) = Z(G)Z(H)$ follows from Theorem 4.38, and it is readily seen that $Z(E;1,\boldsymbol{b},c,1) = c$. $\qquad \square$

Note that a proof of Theorem 4.41 can also be obtained by adapting the proof of Theorem 4.38. This can be done since setting $a = w = 1$ removes the problem that it is not possible to determine the exact relation between $k(H)$ and $k(H/e)$ or between $w(H)$ and $w(H/e)$, when e is a non-trivial orientable loop.

As a corollary, we can obtain a recursive definition of $R(G;x+1,y,1/\sqrt{xy},1)$ by using Proposition 4.37.

Corollary 4.42. *Let G and H be embedded graphs and e be an edge of G. Then*

1. If e is neither a loop nor a bridge

$$R(G;x+1,y,1/\sqrt{xy},1) = R(G-e;x+1,y,1/\sqrt{xy},1) + R(G/e;x+1,y,1/\sqrt{xy},1)$$

2. If e is a bridge

$$R(G;x+1,y,1/\sqrt{xy},1) = (x+1)\cdot R(G-e;x+1,y,1/\sqrt{xy},1)$$

3. If e is a trivial orientable loop

$$R(G;x+1,y,1/\sqrt{xy},1) = (1+y)\cdot R(G-e;x+1,y,1/\sqrt{xy},1)$$

4. If e is a non-trivial orientable loop

$$R(G;x+1,y,1/\sqrt{xy},1)$$
$$= R(G-e;x+1,y,1/\sqrt{xy},1) + (y/x)\cdot R(G/e;x+1,y,1/\sqrt{xy},1)$$

5. If e is a non-orientable loop

$$R(G;x+1,y,1/\sqrt{xy},1)$$
$$= R(G-e;x+1,y,\frac{1}{\sqrt{xy}},1) + \sqrt{\frac{y}{x}}\,R(G/e;x+1,y,1/\sqrt{xy},1)$$

6. $R(G\sqcup H;x+1,y,1/\sqrt{xy},1) = R(G;x+1,y,1/\sqrt{xy},1)R(H;x+1,y,1/\sqrt{xy},1)$

7. $R(E;x+1,y,1/\sqrt{xy},1) = 1$, *when* $E = (\{v\},\emptyset)$

Proof. The result follows from Theorem 4.39 in all cases except when e is a non-trivial orientable loop. Suppose e is a non-trivial loop. Then Eq. (4.17) and Theorem 4.41 give

$$x^{k(G)}(\sqrt{y/x})^{v(G)}R(G;x+1,y,1/\sqrt{xy},1)$$
$$= x^{k(G-e)}(\sqrt{y/x})^{v(G-e)}R(G-e;x+1,y,1/\sqrt{xy},1)$$
$$+(\sqrt{y/x})x^{k(G/e)}(\sqrt{y/x})^{v(G/e)}R(G/e;x+1,y,1/\sqrt{xy},1).$$

Thus,

$$R(G;x+1,y,1/\sqrt{xy},1)$$
$$= x^{k(G-e)-k(G)}(\sqrt{y/x})^{v(G-e)-v(G)}R(G-e;x+1,y,1/\sqrt{xy},1)$$
$$+x^{k(G/e)-k(G)}(\sqrt{y/x})^{v(G/e)-v(G)+1}R(G/e;x+1,y,1/\sqrt{xy},1).$$

Table 4.2 shows the values of the normalisation factors. Substituting these values into the above equation gives the result. $\qquad\square$

4.5.3 Duality Relations for Topological Tutte Polynomials

The ribbon group action on the topological transition polynomial provides a framework for obtaining duality and partial duality relations (see [5, 16, 34, 75, 96]) for the polynomials $Z(G)$ and $R(G)$.

Let G be an embedded graph and let G_m be its embedded medial graph equipped with the canonical checkerboard colouring. Then, for $A \subseteq E(G)$, the weight system $W_Z^{\delta(A)}(G_m)$ is given by reversing the roles of b_e and 1 whenever e is in A, thus:

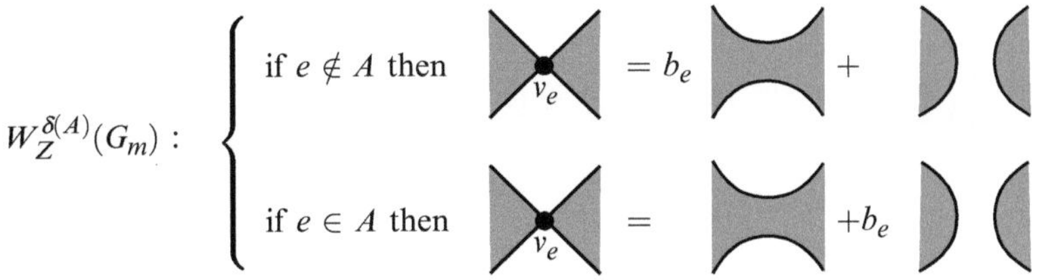

$$W_Z^{\delta(A)}(G_m):\begin{cases}\text{if } e \notin A \text{ then} \quad \times_{v_e} = b_e \;)(\; + \;)(\\[2em] \text{if } e \in A \text{ then} \quad \times_{v_e} = \;)(\; + b_e \;)(\end{cases}$$

Lemma 4.43. *Let G be an embedded graph with embedded medial graph G_m. Then if $A \subseteq E(G)$, we have*

$$Q\big(G; (\boldsymbol{b}, \mathbf{1}, \mathbf{0})^{\delta(A)}, c\big) = \Big(\prod_{e \in A} b_e\Big) Z(G; 1, \boldsymbol{b}_A, c, 1),$$

where $\boldsymbol{b}_A := \{b_e \mid e \notin A\} \cup \{1/b_e \mid e \in A\}$.

Proof. We have

$$Q\big(G; (\boldsymbol{b}, \mathbf{1}, \mathbf{0})^{\delta(A)}, c\big) = \sum_{s \in \mathscr{L}(G_m)} \omega_Z^{\delta(A)}(s)\, c^{c(s)}$$

$$= \sum_{s \in \mathscr{L}(G_m)} \Big(\prod_{\substack{v_e \in Wh(s) \\ e \notin A}} b_e\Big)\Big(\prod_{\substack{v_e \notin Wh(s) \\ e \in A}} b_e\Big) c^{c(s)}, \qquad (4.25)$$

where the sum is over all graph states s with no crossing states, and $Wh(s)$ is the set of vertices with white smoothing states in the graph state s.

We can define a bijection between the set of embedded spanning subgraphs of G and the set of graph states of G_m by associating a spanning subgraph H_s of G to a graph state of G_m by setting $e \in H_s$ if and only if the vertex $v_e \in Wh(s)$. It is then clear that for every graph state, $c(s) = f(H_s)$. By using this bijection, we can write the right-hand side of Eq. (4.25) as

$$\Big(\prod_{e \in A} b_e\Big) \sum_{B \subseteq E(G)} \Big(\prod_{\substack{e \in B \\ e \notin A}} b_e\Big)\Big(\prod_{\substack{e \in B \\ e \in A}} 1/b_e\Big) c^{f(H)} = \Big(\prod_{e \in A} b_e\Big) Z(G; 1, \mathbf{b}_A, c, 1),$$

where $\boldsymbol{b}_A := \{b_e \mid e \notin A\} \cup \{1/b_e \mid e \in A\}$ as required. $\qquad\square$

The partial duality relation for $Z(G)$ given below is an extension of the duality relations in [34, 75].

Theorem 4.44. *Let G be an embedded graph with $A \subseteq E(G)$. Then*

$$Z(G; 1, \boldsymbol{b}, c, 1) = \Big(\prod_{e \in A} b_e \Big) Z(G^{\delta(A)}; 1, \boldsymbol{b}_A, c, 1),$$

where $\boldsymbol{b} = \{b_e | e \in E(G)\}$ *and* $\boldsymbol{b}_A = \{b_e \mid e \notin A\} \cup \{1/b_e \mid e \in A\}$.

Proof. We have

$$\Big(\prod_{e \in A} b_e \Big) Z(G^{\delta(A)}; 1, \boldsymbol{b}_A, c, 1) = Q(G^{\delta(A)}; (\boldsymbol{b}, \boldsymbol{1}, \boldsymbol{0})^{\delta(A)}, c)$$

$$= Q(G; (\boldsymbol{b}, \boldsymbol{1}, \boldsymbol{0}), c) = Z(G; 1, \boldsymbol{b}, c, 1),$$

where the first equality is by Lemma 4.43, the second is by Theorem 4.12, and the third follows from Proposition 4.40. $\qquad\square$

Theorem 4.44 specialises to give a duality relation for $R(G)$.

Theorem 4.45 ([33, 34, 75]). *Let G be an embedded graph. Then*

$$x^{\gamma(G)} R(G; x+1, y, 1/\sqrt{xy}, 1) = y^{\gamma(G)} R(G^*; y+1, x, 1/\sqrt{xy}, 1).$$

Proof. The theorem follows from Eq. (4.18) and Theorem 4.44, by taking $A = E(G)$. $\qquad\square$

4.5.4   *Polynomials of Signed Embedded Graphs*

In applications of graph theory to knot theory, signed graphs and polynomials of signed graphs, including signed versions of topological graph polynomials, arise. These applications will be one of the main topics of Chap. 5. To prepare for this, we now give an overview of a signed version of the ribbon graph polynomial, which is just a special form of the topochromatic polynomial.

A *signed (abstract or embedded) graph* is an abstract or embedded graph whose edges are labelled with elements of $\{+, -\}$. If G is a signed graph, then $E_+(G) \subseteq E(G)$ denotes the set of $+$ weighted edges of G and $e_+(G) := |E_+(G)|$ is the number of such edges. If $A \subseteq E(G)$, then we will use $e_+(A)$ to denote the number of $+$ edges in A. Also, $E_-(G)$, $e_-(G)$, and $e_-(A)$ are defined similarly.

We emphasise that the signs $+$ and $-$ in a signed graph are simply edge weights. Unlike the signs of edges in a signed rotation system, the signs of a signed graph do not record any of the topological properties of an embedded graph. Because of the possibility of confusion, we avoid using signed rotation systems in the same context as signed graphs. We will use signs in signed graphs in Chap. 5 to record the crossing structure of link diagrams and will generally use ribbon graphs to represent embedded graphs when doing so.

Let G be a signed embedded graph and $A \subseteq E(G)$. Then we define the partial dual, $G^{\delta(A)}$, of a signed embedded graph by taking the partial dual of the underlying embedded graph and then assigning signs to the edges as follows so that $G^{\delta(A)}$ is also a signed graph: if an edge e of G has weight $\varepsilon \in \{+, -\}$, then the corresponding edge of $G^{\delta(A)}$ has weight

- $-\varepsilon$ if $e \in A$
- ε if $e \notin A$

Thus, partial duality toggles the signs of the edges in A. We highlight the special case that the geometric dual G^* of a signed graph G is formed by taking the geometric dual of the underlying unsigned graph of G, and then whenever e is an edge of G with sign $+$ or $-$, the sign of e in G^* is $-$ or $+$, respectively.

The signed ribbon graph polynomial was introduced by Chmutov and Pak in [17]. Along with its generalisations it has appeared in several places in the literature (for example [16, 18, 43, 70, 75, 77]). It is defined by the following state sum.

Definition 4.46 (Chmutov and Pak [17]). The *signed ribbon graph polynomial*, $R_s(G;x,y,z) \in \mathbb{Z}[x^{\pm\frac{1}{2}}, y^{\pm\frac{1}{2}}, z^{\pm1}]$, of a signed graph G is defined as

$$R_s(G;x,y,z) = \sum_{A \subseteq E(G)} x^{r(G)-r(A)+s(A)} y^{n(A)-s(A)} z^{k(A)-f(A)+n(A)}, \qquad (4.26)$$

where $s(A) = \frac{1}{2}(e_-(A) - e_-(G-A))$.

We will now see that $R_s(G)$ is a special case of the topochromatic polynomial $Z(G)$. We then use this to determine Chmutov's partial duality relation for the signed ribbon graph polynomial.

Proposition 4.47.

$$R_s\left(G; \frac{ac}{b}, bc, \frac{1}{c}\right) = \left(\frac{b}{ac}\right)^{k(G)} \left(\frac{1}{b}\right)^{v(G)} \left(\frac{b}{\sqrt{a}}\right)^{e_-(G)} Z(G;a,\omega,c,1), \qquad (4.27)$$

where

$$\omega_e = \begin{cases} b & \text{if } e \text{ has positive sign,} \\ \frac{a}{b} & \text{if } e \text{ has negative sign.} \end{cases}$$

Proof. Expanding the rank and nullity in Eq. (4.26) and collecting terms gives

$$R_s(G; x,y,z) = x^{-k(G)} (yz)^{-v(G)} \sum_{A \subseteq E(G)} (xyz^2)^{k(A)} (yz)^{e(A)} z^{-f(A)} (xy^{-1})^{s(A)}.$$

Making the substitutions $a = xyz^2$, $b = zy$, and $c = z^{-1}$ then gives

$$R_s\left(G; \frac{ac}{b}, bc, \frac{1}{c}\right) = \left(\frac{b}{ac}\right)^{k(G)} \left(\frac{1}{b}\right)^{v(G)} \sum_{A \subseteq E(G)} a^{k(A)} b^{e(A)} \left(\frac{a}{b^2}\right)^{s(A)} c^{f(A)}.$$

We now turn our attention to rewriting the sign function $s(A)$ in this expression. The term $e_-(G-A)$ used in the definition of $s(A)$ can be expressed as $e_-(G-A) = e(G) - e_+(G) - e_-(A)$. Substituting this into the formula for $s(A)$ gives

$$s(A) = \frac{1}{2}[e_-(A) - e(G) + e_+(G) + e_-(A)] = \frac{1}{2}[2(e_-(A)) - e_-(G)],$$

where the second equality follows since $e(G) - e_+(G) = e_-(G)$. Thus we have

$$b^{e(A)}\left(\frac{a}{b^2}\right)^{s(A)} = b^{e(A)}\left(\frac{a}{b^2}\right)^{e_-(A) - \frac{1}{2}(e_-(G))}$$

$$= \left(\frac{b}{a^{1/2}}\right)^{e_-(G)}\left[a^{e_-(A)}b^{e(A) - 2(e_-(A))}\right]$$

$$= \left(\frac{b}{a^{1/2}}\right)^{e_-(G)}\prod_{e\in A}\omega_e,$$

since

$$\omega_e = \begin{cases} b & \text{if } e \text{ has positive sign,} \\ \frac{a}{b} & \text{if } e \text{ has negative sign.} \end{cases}$$

We can now express the signed ribbon graph polynomial in terms of Z:

$$R_s\left(G; \frac{ac}{b}, bc, \frac{1}{c}\right) = \left(\frac{b}{ac}\right)^{k(G)}\left(\frac{1}{b}\right)^{v(G)}\left(\frac{b}{\sqrt{a}}\right)^{e_-(G)}\sum_{A\subseteq E(G)} a^{k(A)}c^{f(A)}\prod_{e\in A}\omega_e.$$

$$= \left(\frac{b}{ac}\right)^{k(G)}\left(\frac{1}{b}\right)^{v(G)}\left(\frac{b}{\sqrt{a}}\right)^{e_-(G)} Z(G; a, \omega, c, 1). \qquad \square$$

With the connection between $Z(G)$ and $R_s(G)$ from Proposition 4.47, we see that Theorem 4.44 gives Chmutov's partial duality relation for the signed ribbon graph polynomial.

Theorem 4.48 (Chmutov [16]). *If G is a signed ribbon graph and $G^{\delta(A)}$ is a partial dual of G, then when $xyz^2 = 1$,*

$$(yz)^{v(G)}R_s(G; x, y, z) = (yz)^{v(G^{\delta(A)})}R_s(G^{\delta(A)}; x, y, z). \qquad (4.28)$$

We note that in [96] Vignes-Tourneret defined a multivariate version of the signed ribbon graph polynomial and showed that it satisfies a partial duality relation. It can be shown that this multivariate signed Bollobás–Riordan polynomial is a reformulation of the topochromatic polynomial Z and that the partial duality property of Z_s that was given in [96] also is a consequence of Theorem 4.44.

4.6   Relating the Penrose and Topochromatic Polynomials

The relation between the topochromatic polynomial and the transition polynomial on the one hand, and between the Penrose polynomial and the transition polynomial on the other hand, allows us to obtain a relationship between $P(G)$ and $Z(G)$ (and hence also $R(G)$). We use this to find combinatorial interpretations of evaluations of $Z(G)$ and to reformulate the Four Colour Theorem in terms of $Z(G)$ and $R(G)$.

Note that although in this section we work in terms of the topochromatic polynomial $Z(G)$ (since the resulting formulas are simpler) we can reformulate everything in terms of the ribbon graph polynomial via Proposition 4.37.

Theorem 4.49. *Let G be an embedded graph. Then*

$$P(G;\lambda) = Z(G^{\tau\delta(E)};1,-1,\lambda,1),$$

or equivalently,

$$P(G^*;\lambda) = Z(G^{\times};1,-1,\lambda,1).$$

Proof. We have

$$P(G;\lambda) = Q(G;(1,0,-1),\lambda)$$
$$= Q\left(G^{\tau\delta(E(G))};(-1,1,0),\lambda\right)$$
$$= Z(G^{\tau\delta(E(G))};1,-1,\lambda,1),$$

where the first equality follows from Proposition 4.16, the second follows from Theorem 4.12, and the third equality from Proposition 4.40. The second form then follows by applying this to $G^* = G^{\delta(E(G))}$. $\square$

In Theorem 4.26, we saw that when G is orientable and checkerboard colourable, $P(G;-n) = (-1)^{f(G)} \sum 2^{m(s)}$, where the sum is over all permissible n-valuations s of G_m and $m(s)$ is the number of total vertices in s. We will now extend this result to a larger class of embedded graphs by using Theorem 4.49 and a combinatorial evaluation of the ribbon graph polynomial $R(G)$ that is due to M. Korn and I. Pak. To describe the result we need to define another special type of k-valuation (Definition 4.23).

Definition 4.50. We say that a set of k-valuations of G_m is *z-permissible* if the configuration at each vertex has one of the following forms, where $i \neq j$. Recall that the k-valuation of the last type is called *total*.

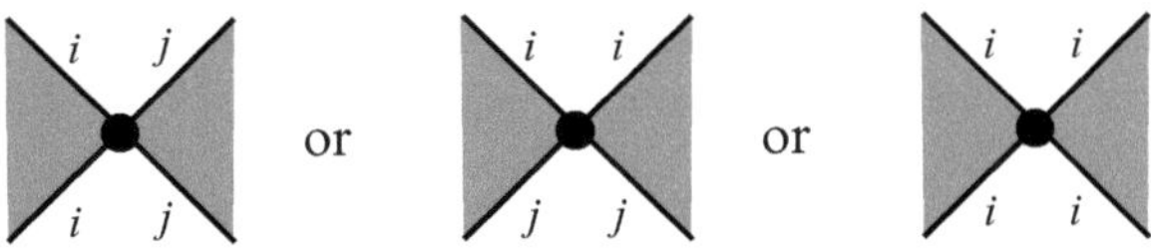

Korn and Pak found the following interpretation of the ribbon graph polynomial, which we state without proof.

Theorem 4.51 (Korn and Pak [61]). *Let G be an embedded graph. Then*

$$k^{k(G)}R(G;k+1,k,1/k,1) = \sum 2^{m(\sigma)},$$

where the sum is over all z-permissible k-valuations σ of G_m, and $m(\sigma)$ is the number of total vertices in σ.

Note that the $k(G)$ in the exponent in Theorem 4.51 is the number of components of G, while the k in the base is the number of colours in the k-valuation.

We can use Theorems 4.49 and 4.51 to obtain the following combinatorial interpretation of $P(G;k)$ that extends Theorem 4.26.

Theorem 4.52. *Let G be an embedded graph such that $(G^*)^\times$ is orientable. Then $P(G;-k) = (-1)^{f(G)}\sum 2^{m(s)}$, where the sum is over all permissible k-valuations s of G_m and $m(s)$ is the number of total vertices in s.*

Proof. By Eq. (4.18),

$$Z(G;1,-1,-k,1) = k^{k(G)}(-1)^{v(G)}R(G;k+1,k,-1/k,1).$$

If G is orientable, then the c's in $R(G;a,b,c,w)$ always appear with even exponents. Thus if G is orientable,

$$Z(G;1,-1,-k,1) = (-1)^{v(G)}k^{k(G)}R(G;k+1,k,-1/k,1)$$
$$= (-1)^{v(G)}k^{k(G)}R(G;k+1,k,1/k,1)$$
$$= (-1)^{v(G)}\sum_{\sigma\in\mathscr{Z}_k(G_m)} 2^{m(\sigma)},$$

where $\mathscr{Z}_k(G_m)$ is the set of permissible k-valuations of G_m. Using Theorem 4.49, when $(G^*)^\times$ is orientable, we have

$$P(G;-k) = Z(G^*)^\times;1,-1,-k,1)$$
$$= (-1)^{v(G^*)^\times}\sum_{\sigma\in\mathscr{Z}_k(((G^*)^\times)_m)} 2^{m(\sigma)}$$
$$= (-1)^{f(G)}\sum_{\sigma\in\mathscr{Z}_k(((G^*)^\times)_m)} 2^{m(\sigma)},$$

where the third equality follows since $(G^*)^\times = v(G^*) = f(G)$.

To rewrite $\mathscr{Z}_k(((G^*)^\times)_m)$ in terms of permissible k-valuations of G, consider Fig. 4.8. An edge e in the arrow presentations of G and $(G^*)^\times$ is shown in Fig. 4.8a and b, respectively. The top row of the table in Fig. 4.8c shows the vertex v_e and

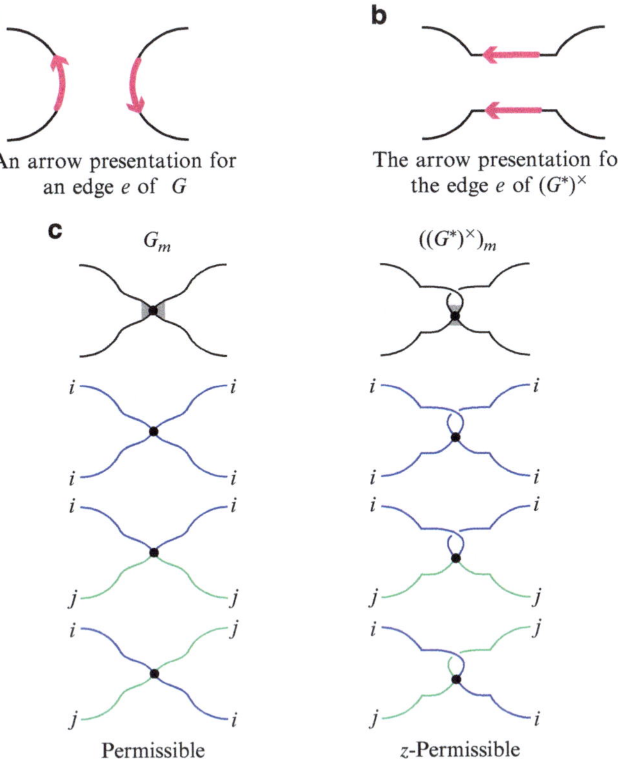

A table showing G_m at v_e, and its permissible k-valuations;
and $((G^*)^\times)_m$ at v_e, and its z-permissible k-valuations

Fig. 4.8 Identifying the permissible k-valuations of G_m and the z-permissible and k-valuations of $((G^*)^\times)_m$

boundary components of G_m and $(G^*)^\times{}_m$, with the shading indicating the canonical checkerboard colouring at v_e. The first column shows, at v_e, the permissible k-valuations of G_m and the second column shows the k-valuations of $((G^*)^\times)_m$ in $\mathscr{L}_k(((G^*)^\times)_m)$. We see from the table that there is a bijection between the permissible k-valuations of G_m and the k-valuations of $(G^*)^\times{}_m$ in $\mathscr{L}_k(((G^*)^\times)_m)$. Moreover, this bijection preserves the number of total vertices in the k-valuations. Thus,

$$\sum_{\sigma \in \mathscr{L}_k(((G^*)^\times)_m)} 2^{m(\sigma)} = \sum 2^{m(s)},$$

where the second sum is over all permissible k-valuations s of G_m and $m(s)$ is the number of total vertices in s. The result then follows. $\square$

To see that Theorem 4.52 genuinely does generalise Theorem 4.26, observe that $(G^*)^\times$ being orientable is weaker than G being orientable and checkerboard colourable. For example, let G be the embedded graph consisting of one vertex and three non-orientable loops that meet the vertex in the cyclic order $(abcabc)$. Then $(G^*)^\times$ is orientable (it is a plane 3-cycle), but G is neither orientable nor checkerboard colourable. On the other hand, if G is orientable and checkerboard colourable, then $(G^*)^\times$ is orientable and bipartite, so $(G^*)^\times$ is orientable.

Theorem 4.52 illustrates how the connection between the graph polynomials P and Z from Theorem 4.49 can be used to obtain new results about the Penrose polynomial. Similarly, Theorem 4.49 can be used in the other direction to obtain new results about the ribbon graph polynomial. Again, although we work in terms of $Z(G)$, the results can be rewritten in terms of $R(G)$ using Eq. (4.18).

Theorem 4.53. *Let G be an embedded graph.*

1. If $H := (G^\times)^*$ is plane, connected, and cubic, then the number of edge 3-colourings of H is

$$Z(G; 1, -\mathbf{1}, 3, 1) = (-1/4)^{v(H)/2} Z(G; 1, -\mathbf{1}, -2, 1).$$

2. We have
$$Z(G; 1, -\mathbf{1}, \lambda, 1) = \sum_{A \subseteq E(H)} (-1)^{|A|} \chi((G^{\delta(A)})^\times; \lambda).$$

Proof. Item 1 follows from Theorem 4.49 and Eq. (4.8).

For Item 2, observe that Theorem 4.49 and Theorem 4.28 give

$$Z(G; 1, -\mathbf{1}, \lambda, 1) = \sum_{A \subseteq E((G^\times)^*)} (-1)^{|A|} \chi(((G^\times)^*)^{\tau(A)})^*; \lambda).$$

Rewriting the exponent, with $A^c := E(G) \backslash A$, gives

$$\delta\tau(E(G))\tau(A)\delta(E(G)) = \delta\tau\delta\tau(A)\delta\delta\tau(A^c) = \tau\delta(A)\tau(A^c) = (G^{\delta(A)})^\times,$$

and the result follows. $\square$

We can also use the relation between the Penrose and topological Tutte polynomials to obtain a new formulation of the Four Colour Theorem in terms of $Z(G)$ or $R(G)$.

Corollary 4.54. *The following statements are equivalent:*

1. The Four Colour Theorem is true.
2. For every connected, loopless plane graph G, $Z(G^\times; 1, -\mathbf{1}, 3, 1) > 0$, or equivalently, $(-1)^{v(G)} R(G^\times; -2, -3, 1/3, 1) < 0$.
3. For every connected, loopless plane graph G, $Z(G^\times; 1, -\mathbf{1}, 4, 1) > 0$, or equivalently $(-1)^{v(G)} R(G^\times; -3, -4, 1/4, 1) < 0$.
4. For every connected, loopless plane graph G, $Z(G^\times; 1, -\mathbf{1}, -2, 1) \neq 0$, or equivalently, $R(G^\times; 3, 2, -1/2, 1) \neq 0$.

Proof. Corollary 9 of [2] states that the Four Colour Theorem is equivalent to $P(G;3) > 0$ for all connected, bridgeless plane graphs G. This is equivalent to $P(G^*;3) > 0$ for all connected, loopless plane graphs G, which, by Theorem 4.49, is equivalent to $Z(G^\times;1,-1,3,1) > 0$ for all connected, loopless plane graphs G. Translating to $R(G)$ using Eq. (4.18) gives

$$(-1)^{r(G^\times)}3^{k(G^\times)}R(G^\times;-2,-3,1/3,1) > 0,$$

or, since G is connected and has the same rank and number of vertices as $G^\times$, this is equivalent to $(-1)^{v(G)}R(G^\times;-2,-3,1/3,1) < 0$, which yields the rest of the result.

The next item is shown in a similar way but using the fact from [2] that the Four Colour Theorem is equivalent to showing that $P(G;4) > 0$ for all connected, bridgeless plane graphs G.

The final item follows from [49], Proposition [7] which says that if G is plane and connected, then $(-1)^{e(G)}(-1/2)^{v(G)-2}P(G;-2)$ is equal to the number of face 4-colourings of G $\square$

Chapter 5
Applications to Knot Theory

There is a classical and well-known way to associate signed plane graphs, called (signed) Tait graphs, to link diagrams. These are formed in the same way as Tait graphs of 4-regular graphs, after viewing the link diagram as an embedded 4-regular graph, but with signs on the edges to record the crossing structure of the link diagram (see Sect. 5.3.1). Signed Tait graphs are a standard tool in knot theory. They provide a bridge between knot theory and graph theory and have found numerous applications in both areas. Recently, in [21], Dasbach et al. (see also Turaev [95]) extended the idea of a signed Tait graph by associating a set of embedded graphs to a link diagram. In their construction, each embedded graph arises by assigning one of the two possible smoothings at each crossing of the link diagram. The signed Tait graphs of a link diagram appear in this set of embedded graphs. One of the key advantages to this approach of using non-plane graphs to describe links is that it provides a way to encode the crossing structure of a link diagram in the topology of the embedded graph, thus avoiding signs on the edges of an embedded graph. This idea has found many recent applications in knot theory, such as the Jones and HOMFLY-PT polynomials [16–18, 21, 76, 77, 97], Khovanov homology [14], knot Floer homology [71], Turaev genus [1, 71, 95], quasi-alternating links [99], the coloured Jones polynomial [39], the signature of a knot [23], the determinant of a knot [21, 22], and hyperbolic knot theory [38].

In this chapter we survey some of these recent connections between knot theory and graph theory. After giving a quick overview of classical links, links in thickened surfaces and virtual links, we explain some of the ways in which embedded graphs can be used to describe link diagrams. We go on to describe recently discovered connections between the ribbon graph polynomial and the Jones and HOMFLY-PT polynomials of a link. For classical links, these give connections between the Tutte polynomial and knot polynomials. As the intent of this chapter is to survey connections between partial duality, graph polynomials, and knot theory, rather than to be a complete treatise on knot theory, we assume a familiarity with basic knot theory and at times refer the reader to the literature for details of well-established results.

J.A. Ellis-Monaghan and I. Moffatt, *Graphs on Surfaces: Dualities, Polynomials, and Knots*, SpringerBriefs in Mathematics, DOI 10.1007/978-1-4614-6971-1_5,
© Joanna A. Ellis-Monaghan, Iain Moffatt 2013

We generally work in the language of ribbon graphs, which is the most convenient realisation of embedded graphs for our purposes here.

5.1 Knots and Links

In this section we briefly review the fundamentals of knots and links in the 3-sphere and thickened surfaces $\Sigma \times I$ and their diagrams. We assume a familiarity with the basics of classical knot theory (i.e., of knots and links in S^3 or $\mathbb{R}^3$) and indicate how the theory extends to knots and links in thickened surfaces. We omit the proofs of many standard results, referring the reader to [68] or any other of the many excellent introductory texts on knot theory for further details.

5.1.1 Links in a 3-Manifold

An n-component *link* in a 3-manifold M is a subset of M that consists of n disjoint piecewise-linear simple closed curves. A *knot* is a one-component link. Figure 5.1 shows some knots and links which have been thickened to better illustrate their structure. A link is said to be *oriented* if each of its components is oriented. (An orientation is just a preferred direction around each component and is usually indicated by arrows on the link.) A knot is said to be the *unknot* if it bounds a disc embedded in M. A link is the *unlink* if it is the boundary of a set of discs embedded in M.

The piecewise-linear condition (which means that links are made up of a finite number of straight line segments) in the definition of a link is included in order to avoid pathologies. (The more general definition of a knot as embedding of S^1 into M results in a theory with very unexpected features; see, for example, [9,35,36,74,88].) As is usual, when we draw links, we will assume that they are made of a sufficiently large number of sufficiently small line segments so that they appear to be smooth.

Fig. 5.1 Knots and links in S^3

Here we are only interested in links in certain 3-manifolds. We will take M to be either the 3-sphere, S^3, or a thickened surface $\Sigma \times I$, where Σ is a closed orientable surface and I is the unit interval $[0,1]$. A link in the 3-sphere is a *classical link*, and a link in $\Sigma \times I$ is a *link in a thickened surface*. We will always assume the 3-manifold M is oriented.

Intuitively, we consider two links $L, L' \subset M$ to be equivalent if we can deform the manifold M in such a way that one link is taken to the other. Formally, two links L and L' in M are *equivalent* if there is an orientation preserving piecewise-linear homeomorphism $h : M \to M$ such that $h(L) = L'$. (The piecewise-linear condition on the homeomorphism means that, possibly after subdividing simplices in M, h sends simplices to simplices.) Equivalent links are also said to be ambient isotopic. See [9] for details on isotopy and various definitions of link equivalence.

A function f defined on links is a *knot invariant* if $f(L) = f(L')$ whenever L and L' are equivalent. Note that $f(L) = f(L')$ does not necessarily imply that $L = L'$.

Knot theory is usually concerned with the equivalence classes of links. With this in mind, we note that knot theory in S^3 is equivalent to knot theory in $\mathbb{R}^3$ and is also equivalent to knot theory in $S^2 \times I$.

5.1.2 Link Diagrams

Rather than working directly with links in S^3 or $\Sigma \times I$, we will work with link diagrams. A diagram D of a link L is essentially a drawing of it on a surface (either on S^2 for links in S^3 or on Σ for links in $\Sigma \times I$). Suppose L is a link in S^3. View S^3 as $\mathbb{R}^3 \cup \infty$ and S^2 as the copy of $\mathbb{R}^2 \cup \infty$ given by the xy-plane. Let $p : \mathbb{R}^3 \to \mathbb{R}^2$ be the standard projection. The image $p(L)$ is a *link projection*. By deforming L slightly if required, we can assume that $p(L)$ contains finitely many multiple points, and at each multiple point exactly two arcs meet transversally (see [19]). Such a projection is said to be *regular*. A *diagram* of L consists of a regular projection of L equipped with a crossing structure at each double point, as in Fig. 5.2b. The surface S^2 on which the link diagram is drawn is oriented.

Conversely, given a link diagram D on S^2, a link can be obtained by regarding S^2 as a subset of S^3 and "pulling the overcrossing up a little" into the interior of S^3.

Similarly, if L is a link in $\Sigma \times I$, then a *diagram* on Σ for L can be obtained by projecting $\Sigma \times I$ onto $\Sigma \times \{0\}$ by the mapping $(p,q) \mapsto (p,0)$ (deforming L slightly, if needed, so that this projection is regular, that is it contains finitely many multiple points and the arcs meet transversally at them) and equipping the resulting image of L with a crossing structure in the natural way, as in Fig. 5.2. The surface Σ is oriented with its orientation inherited from $\Sigma \times I$.

Again, given a link diagram D on Σ, a link can be obtained by regarding Σ as $\Sigma \times \{0\} \subset \Sigma \times I$ and "pulling the overcrossing up a little" into the interior of $\Sigma \times I$.

A link diagram D on S^2 or Σ is said to be *connected* if it defines a connected subspace of S^2 or Σ (i.e., there is a path between any two points in $D \cap S^2$ or $D \cap \Sigma$). The faces of D are the connected components of $S^2 \backslash D$ or $\Sigma \backslash D$. The faces of a

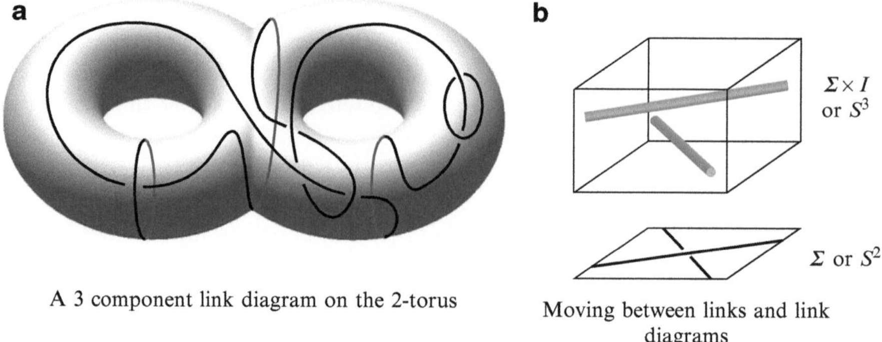

A 3 component link diagram on the 2-torus

Moving between links and link diagrams

Fig. 5.2 Link diagrams

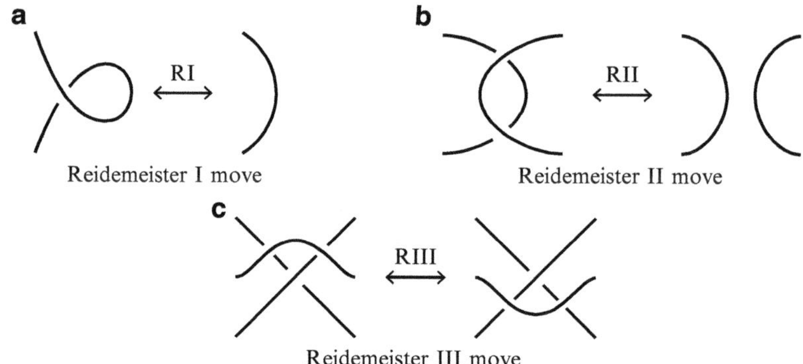

Reidemeister I move

Reidemeister II move

Reidemeister III move

Fig. 5.3 The Reidemeister moves

link diagram need not be discs. Note that classical link diagrams and link diagram on surfaces can be regarded as 4-regular graphs, with crossing information at the vertices, and possibly with some additional closed loop components.

Links in S^3 and $\Sigma \times I$ are considered up to equivalence. The moves on a link diagram corresponding to link equivalence are the Reidemeister moves (which can change the crossing structure) and orientation preserving homeomorphisms of the surface (which allow us to deform a link diagram by "pushing the surface around"). A *Reidemeister move* is a move which changes a link diagram in one of the ways shown in Fig. 5.3. The link diagrams are unchanged outside of the regions shown. The Reidemeister moves are of three types, I, II, and III, as indicated.

We will say that two link diagrams (on S^2 or Σ) are *equivalent* if there is a sequence of Reidemeister moves and orientation preserving homeomorphisms of the surface taking one diagram to the other.

The following equivalence between links and link diagrams is well known. A proof of the result for links in S^3 can be found in, for example, [9] and is readily adapted to links in $\Sigma \times I$.

Theorem 5.1. *Let L and L' be links in S^3 (or $\Sigma \times I$) represented by diagrams D and D' on a surface S^2 (or Σ). Then L and L' are equivalent if and only if D and D' are equivalent.*

Two link diagrams are related by *regular isotopy* if there is a sequence of Reidemeister type II and III moves (see Fig. 5.3b, c) and orientation preserving homeomorphisms of the surface taking one to the other. That is, for regular isotopy, we do not allow the Reidemeister I move.

For oriented links and their oriented link diagrams, the Reidemeister moves preserve the orientations of the arcs they involve. Forming all possible orientations of the arcs in the Reidemeister moves in Fig. 5.3 gives a set of 14 oriented Reidemeister moves. Many of these moves, however, are consequences of the others, and only four oriented Reidemeister moves are actually needed (see [85] for details).

5.2   Virtual Links

Virtual knot theory, introduced by Kauffman in [55], is an extension of classical knot theory. A virtual link diagram is essentially a link diagram in which some of the crossings are designated as virtual crossings (see Fig. 5.4). Virtual link diagrams are considered up to the generalised Reidemeister moves, shown in Fig. 5.5, where the Reidemeister moves are augmented to incorporate virtual crossings. Virtual knot theory can be regarded as the study of links in thickened surfaces through their projections on to the plane, with the caveat that you *do not* know the genus of the surface that the link diagram was drawn on. In this model, the virtual crossings arise as artefacts of the projection. We note that virtual knots can also be regarded as the knotted objects that can be represented by Gauss diagrams (see [40]), but we do not pursue this direction here. Here we give a brief overview of virtual knots, starting with virtual link diagrams, then describing their connections with links in thickened surfaces. We state key results on virtual knots without proof, referring the reader to their sources or to the surveys [55–58, 72] for additional details.

5.2.1   Virtual Link Diagrams

A virtual link diagram is essentially a link diagram on the plane in which some of the crossings are marked as *virtual crossings*. Formally, a *virtual link diagram* consists of n closed piecewise-linear plane curves in which there are finitely many multiple points, and these have the property that at each multiple point exactly two arcs meet and they meet transversally. Moreover, each double point either is assigned a *classical crossing* structure (as in Fig. 5.4a) or is marked as a *virtual crossing* (shown in Fig. 5.4b).

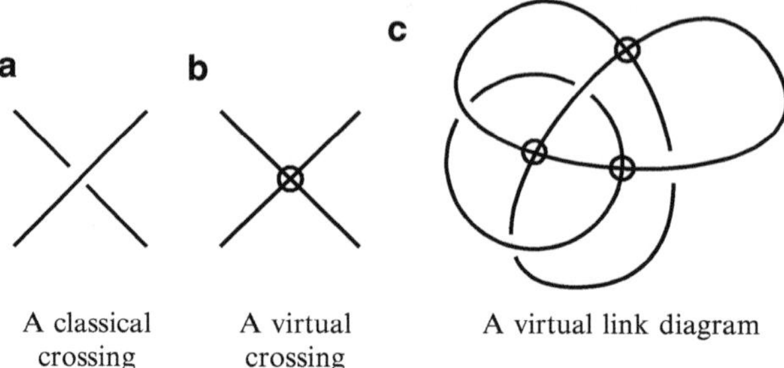

Fig. 5.4 Virtual link diagrams

As with link diagrams, we assume that the piecewise-linear structure is so fine that arcs appear to be smooth. We also emphasise that virtual crossings have no over/undercrossing structure.

Virtual links are considered up to the *generalised Reidemeister moves* shown in Fig. 5.5. These consist of the classical Reidemeister moves of Fig. 5.3 together with the *virtual Reidemeister moves* (which involve only virtual crossings) and the *semivirtual move* (which involves both classical and virtual crossings).

Two virtual link diagrams are *equivalent* if there is a sequence of generalised Reidemeister moves and orientation preserving homeomorphisms of the plane taking one diagram to the other.

A useful consequence of the generalised Reidemeister moves is the *detour move* shown in Fig. 5.6a. Note, however, that the *forbidden moves* shown in Fig. 5.6b are not included in set of the generalised Reidemeister moves. In fact, it was shown in [82] that if forbidden moves are allowed, then all virtual knots become equivalent.

While it might appear that the generalised Reidemeister moves create a new equivalence relation on classical link diagrams, this turns out not to be the case. If D and D' are two classical link diagrams that are related by generalised Reidemeister moves, then they are also related by classical Reidemeister moves:

Theorem 5.2 (Goussarov et al. [40]). *Let D and D' be two classical link diagrams. Then D and D' are equivalent as virtual link diagrams if and only if they are equivalent as classical link diagrams.*

5.2.2   Virtual Links as Links in Thickened Surfaces

Given a virtual link diagram D, we can obtain a link diagram on a surface by drawing D on S^2 by regarding S^2 as $\mathbb{R}^3 \cup \infty$, and then adding handles to S^2, as shown in Fig. 5.7, to eliminate all of the virtual crossings. This gives a link diagram on a surface, which is called a *surface embedding* of D.

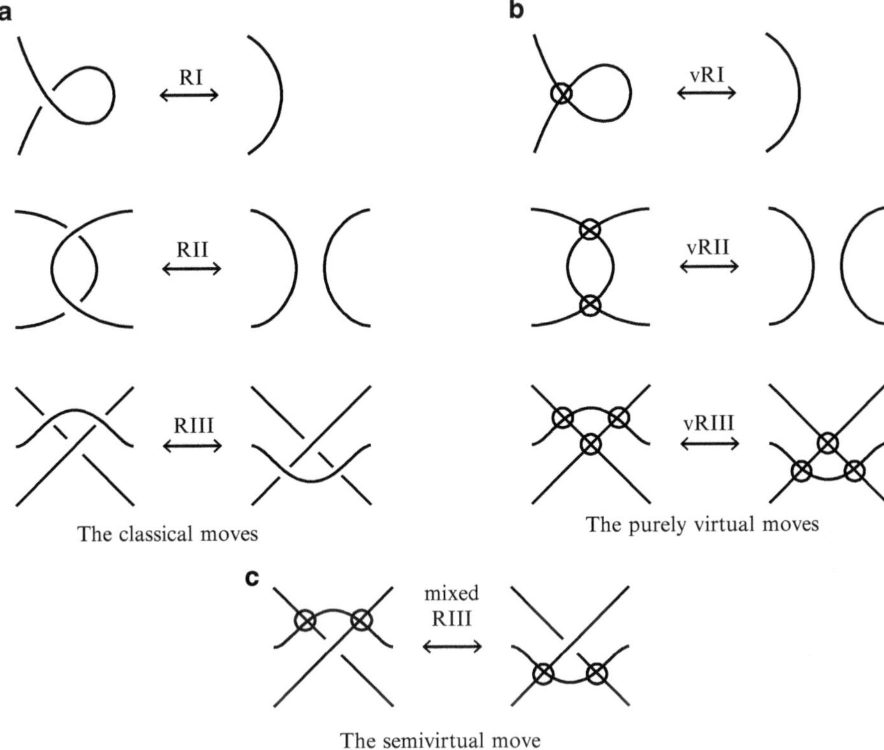

The classical moves

The purely virtual moves

The semivirtual move

Fig. 5.5 The generalised Reidemeister moves

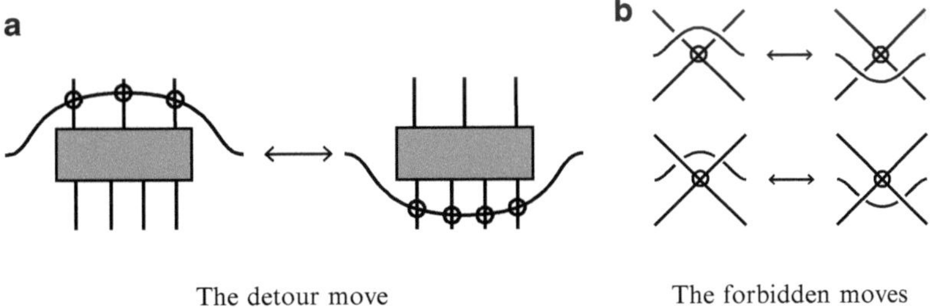

The detour move

The forbidden moves

Fig. 5.6 Further moves on virtual link diagrams

Two surface embeddings of two virtual link diagrams are said to be *stably equivalent* if one can be obtained from the other by the addition or subtraction of handles that do not intersect arcs of the link diagram (so handles can be added to or removed from the surface provided that the link diagram does not pass over them), by the classical Reidemeister moves, and by orientation preserving

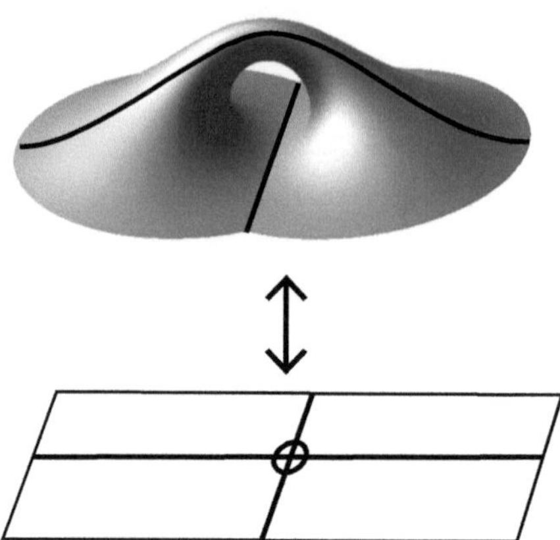

Fig. 5.7 Moving between virtual link diagrams and link diagrams on surfaces

homeomorphisms of the surface. Virtual knot theory is equivalent to knot theory in thickened surfaces up to stable equivalence:

Theorem 5.3 (Kauffman [55]; Carter et al. [12]). *Two virtual link diagrams are equivalent if and only if their corresponding surface embeddings are stably equivalent.*

5.3 Presenting Links as Embedded Graphs

In this section we describe the signed ribbon graphs of a link diagram. Our starting point is the classical construction of signed Tait graphs of link diagrams. Then, by changing from the language of cellularly embedded graphs to arrow presentations, we show how signed Tait graphs can be generalised to give the signed ribbon graphs of a link diagram. The approach here is very similar to that used to introduce and define the cycle family graphs of a medial graph in Chap. 3.

5.3.1 Signed Tait Graphs

Let D be either a classical link diagram or a link diagram on a surface. A *checkerboard colouring* is an assignment of the colour black or white to each face of D in a way such that adjacent faces have different colours. Observe that like 4-regular plane graphs, link diagrams in S^2 are always checkerboard colourable (although a diagram in an arbitrary surface need not be). This is since classical link diagrams and link diagram on surfaces can be regarded as decorated 4-regular graphs, possibly with some additional closed loop components. (We will discuss checkerboard colourings and signed Tait graphs for virtual links in Sect. 5.3.2.)

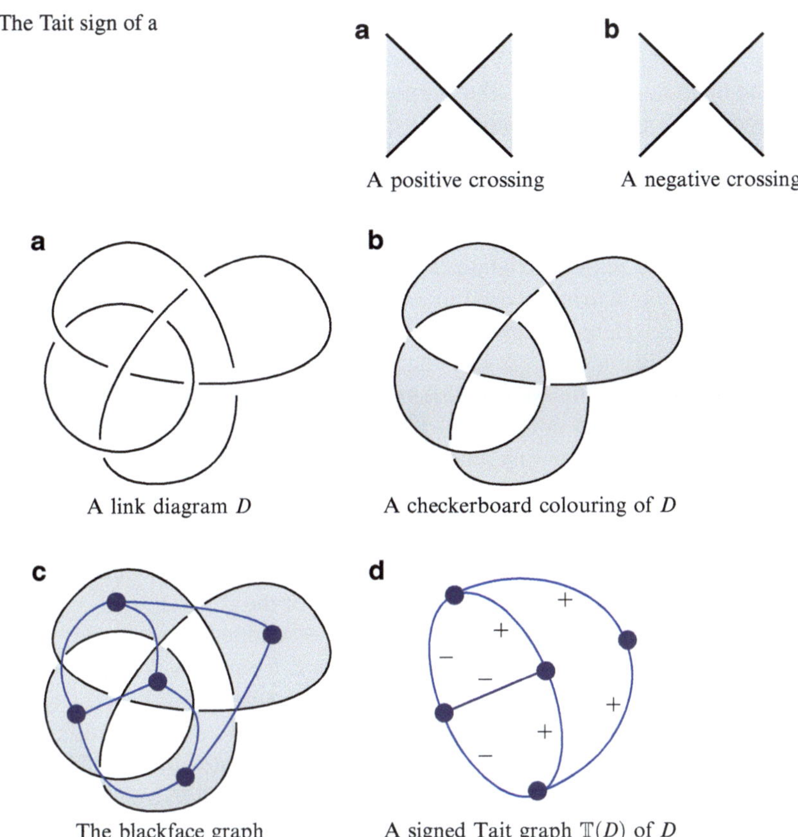

Fig. 5.8 The Tait sign of a crossing

Fig. 5.9 Forming a signed Tait graph of a classical link diagram

The *Tait sign* of a crossing in a checkerboard coloured link diagram is an element of $\{+, -\}$ which is assigned to the crossing as shown in Fig. 5.8.

Let D be a connected checkerboard colourable link diagram on S^2 or Σ. A *signed Tait graph*, $\mathbb{T}(D)$, of D is a signed embedded graph constructed as follows: if D contains crossings checkerboard colour the link diagram; by regarding the link diagram as an embedded graph in which the double points are vertices, form the black face graph; weight each edge of the graph with the Tait sign of the corresponding crossing. If D has no crossings, place a vertex in the black face. An example is given in Fig. 5.9.

Note that by changing crossings to vertices, a checkerboard colourable link diagram gives rise to a checkerboard colourable 4-regular embedded graph. The Tait graphs of both the link diagram and the embedded graph agree (although the Tait graph of the link diagram is signed). This observation justifies the use of the term Tait graph both for embedded graphs and for link diagrams, which should cause no confusion.

If the link diagram $D \subset \Sigma$ is not connected with, say, $D = D_1 \sqcup \cdots \sqcup D_k$, then we have $\Sigma = \Sigma_1 \# \cdots \# \Sigma_k$ with $D_i \subset \Sigma_i$, for each i. The Tait graphs of D are then defined as the disjoint union of the Tait graphs of the D_i. For simplicity, we will generally consider Tait graphs of connected link diagrams. Note that any link diagram can be made into an equivalent connected link diagram using Reidemeister II moves.

Remark 5.4. Signed Tait graphs are classically defined for link diagrams in $\mathbb{R}^2$, in which case canonical checkerboard colourings can be obtained by insisting that the unbounded region is white. Often in the literature this canonical choice of checkerboard colouring is used to associate a unique signed Tait graph with a link diagram on $\mathbb{R}^2$. Here, however, we consider link diagrams on S^2 and Σ and so cannot make such a canonical choice of checkerboard colouring. A link diagram therefore gives rise to two (or no) signed Tait graphs. Accordingly, our use of $\mathbb{T}(D)$ is, strictly speaking, an abuse of notation. However, as we will see shortly, the two signed Tait graphs present the same link diagram, and so the ambiguity in this notation does not cause problems here.

By Item 4 of Theorem 1.15, the two signed Tait graphs associated with a link diagram are geometric duals of one another. (We recall from Sect. 4.5.4 that duality for signed embedded graphs toggles the signs of the edges.)

If $G \subset S^2$ or $G \subset \Sigma$ is a signed cellularly embedded graph, then we can recover a unique link diagram $D(G) \subset S^2$ or $D(G) \subset \Sigma$ from G by forming its canonically checkerboard coloured medial graph G_m, signing each vertex in G_m using the sign of the corresponding edge in G, then assigning a crossing structure according to Fig. 5.8. This process reverses the construction illustrated in Fig. 5.9.

We therefore have that every signed embedded graph in a surface S^2 or Σ is the signed Tait graph of a unique link diagram on S^2 or Σ. In particular, every signed plane graph is the signed Tait graph of a unique classical link diagram. Since the two signed Tait graphs of a link diagram are geometric duals, and duality changes the sign of an edge, they give rise to the same link diagram.

We summarise this discussion and some other basic properties of signed Tait graphs below. The properties follow easily from the results in Sect. 1.5.3 on the Tait graphs of 4-regular embedded graphs. We leave a formal proof of the proposition as an exercise.

Proposition 5.5. *Let D be a connected checkerboard colourable link diagram on a surface S^2 or Σ, and let G be a signed embedded graph and G^* be its signed geometric dual. Then the following properties hold:*

1. $D(\mathbb{T}(D)) = D$
2. $\mathbb{T}(D(G)) \in \{G, G^*\}$
3. $D(G) = D(G^*)$

5.3.2   *Ribbon Graphs and Link Diagrams*

The ribbon graphs of a link diagram were originally defined for classical links by Dasbach et al. in [21] (see also Turaev [95]) and for virtual links by Chmutov and Voltz in [18] and Bradford et al. in [8].

In order to introduce and motivate the construction of the signed ribbon graphs of a link diagram, we consider the construction of signed Tait graphs in terms of signed arrow presentations. Suppose that D is a checkerboard coloured classical link diagram or link diagram on a surface. To obtain a signed arrow presentation of $\mathbb{T}(D)$, start by assigning a label to each crossing of D, then replace each vertex of D with one of the two signed arrow marked consistent smoothings shown in Fig. 5.10 according to the checkerboard colouring. The resulting signed arrow presentation describes $\mathbb{T}(D)$.

Thus, the signed Tait graphs of a link diagram D are formed by using the checkerboard colouring to choose a particular smoothing at each vertex of D. The signed ribbon graphs of a link diagram arise by relaxing this condition. That is, at each vertex we can choose *either* of the two arrow marked smoothings shown in Fig. 5.10. In particular, this means that we no longer require D to be checkerboard colourable, and so ribbon graphs can be obtained for *all* link diagrams in surfaces and for *all* virtual link diagrams, not just checkerboard colourable ones. We now give the formal definition of the ribbon graphs of a link diagram.

Let D be a classical link diagram, a link diagram on a surface, or a virtual link diagram. Assign a unique label to each crossing of D. An *arrow marked A-smoothing* or an *arrow marked B-smoothing* of a crossing c is the replacement of the crossing with one of the two schemes shown in Fig. 5.11. Notice that we decorate the two arcs in the smoothing with signed, labelled arrows, chosen to be both consistent with (or both counter to) the orientation of the surface on which the diagram is drawn. The labels of the arrows are determined by the label of the crossing, and the signs are determined by the choice of smoothing. When rotating around the crossing in the direction of the orientation of the surface, an A-smoothing joins the lower strand to the upper and a B-smoothing joins the upper to the lower. An *arrow marked state*, s, of a link diagram is an assignment of an arrow marked A-smoothing or an arrow marked B-smoothing to each crossing. Observe that an arrow marked state is precisely a signed arrow presentation.

Definition 5.6. Let D be a classical link diagram, a link diagram on a surface, or a virtual link diagram. In addition let s be an arrow marked state of D. Then the signed

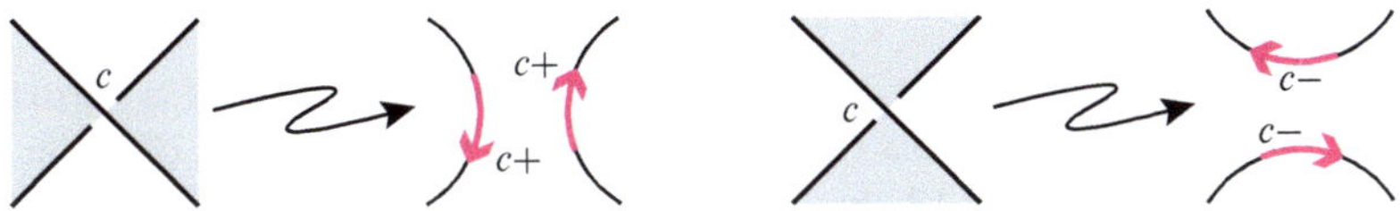

Fig. 5.10   Forming a signed Tait graph using arrow presentations

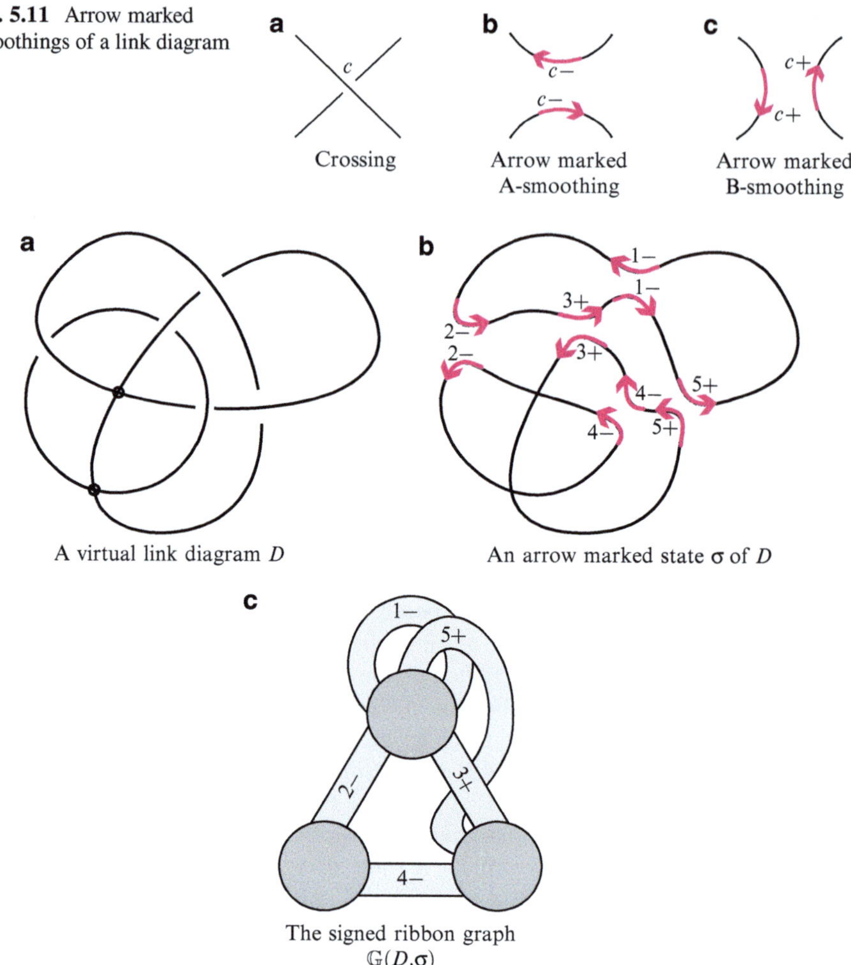

Fig. 5.12 A signed ribbon graph of a virtual link diagram

ribbon graph $\mathbb{G}(D,s)$ corresponding to the signed arrow presentation s is a *signed ribbon graph of the link diagram D*.

The *set of signed ribbon graphs of D*, denoted by $\mathscr{G}(D)$, is defined by

$$\mathscr{G}(D) := \{\mathbb{G}(D,s) \mid s \text{ is an arrow marked state of } D\}.$$

Figure 5.12 provides an example of a signed ribbon graph of a virtual link diagram. Note that, in general, a signed ribbon graph of a virtual link diagram might be non-orientable. Figure 5.13 shows the set of signed ribbon graphs of the trefoil knot.

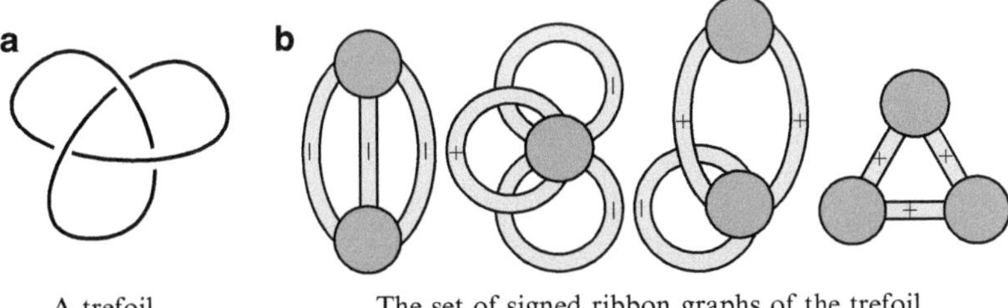

Fig. 5.13 The set of signed ribbon graphs of a trefoil

For comparison with the literature, we note that our sign convention has been chosen to be consistent with the Tait sign of a crossing. However, this sign convention is the opposite from that used by Chmutov in [16].

We will now describe some special elements of $\mathscr{G}(D)$.

Signed Tait Graphs: Suppose D is a checkerboard colourable classical link diagram or link diagram on a surface. Let $\mathbb{T}(D)$ be a signed Tait graph of D. Then $\mathbb{T}(D) \in \mathscr{G}(D)$, since $\mathbb{T}(D) = \mathbb{G}(D,s)$, where the marked state s is obtained by choosing an arrow marked A-smoothing at each $-$ crossing, and an arrow marked B-smoothing at each $+$ crossing, where the signs $+$ and $-$ are the Tait signs of the crossings with respect to a checkerboard colouring of D. In this construction, the arrow marked states are chosen so that the curves follow the black faces of the checkerboard coloured link diagram.

This formulation of signed Tait graphs extends to virtual links. The notion of a checkerboard colouring of a virtual link diagram was introduced by Kamada in [59, 60]. A virtual link diagram is said to be *checkerboard colourable* if there is a colouring of a small neighbourhood of one side of each arc in the diagram such that near a classical crossing the colouring alternates, and near a virtual crossing the colourings go through without noticing the crossing strand and its colouring. An example of a checkerboard coloured virtual link is given in Fig. 5.14a. Note that not every virtual link diagram is checkerboard colourable. Each checkerboard coloured classical crossing has a Tait sign. The *signed Tait graph* $\mathbb{T}(D)$ of a checkerboard colourable virtual link diagram D is defined by $\mathbb{T}(D) = \mathbb{G}(D,s)$, where the state s is obtained by choosing an arrow marked A-smoothing at each $-$ crossing and an arrow marked B-smoothing at each $+$ crossing. Figure 5.14b shows a signed Tait graph of the virtual link diagram of Fig. 5.14a.

The All-A and All-B Signed Ribbon Graphs: The *all-A signed ribbon graph* is defined as $\mathbb{A}(D) := \mathbb{G}(D,s)$, where s is the arrow marked state obtained by choosing an arrow marked A-smoothing at each crossing. See Fig. 5.15b, c for an example (for clarity, in the figure we are using a single label to denote the labels of a pair of arrows).

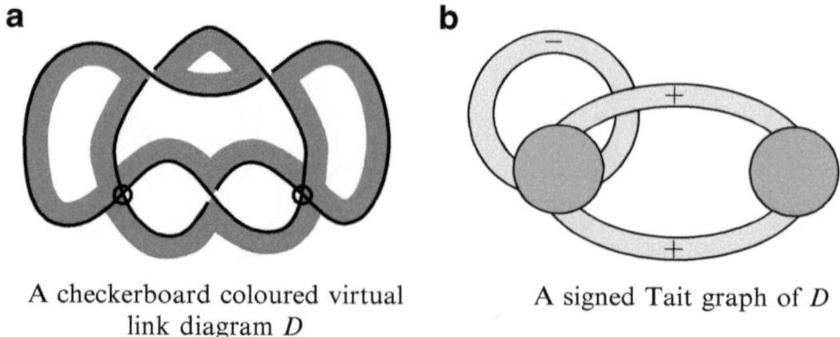

A checkerboard coloured virtual
link diagram D

A signed Tait graph of D

Fig. 5.14 The signed Tait graph of a virtual link diagram

Similarly, the *all-B signed ribbon graph* is defined as $\mathbb{B}(D) := \mathbb{G}(D,s)$, where s is the arrow marked state obtained by choosing an arrow marked B-smoothing at each crossing.

The underlying surfaces of the all-A signed ribbon graphs were previously considered by Turaev in [95] and are known as *Turaev surfaces*. If L is a link, then the minimum genus of $\mathbb{A}(D)$ over all diagrams D presenting L is known as the *Turaev genus* of L.

In the all-A and all-B signed ribbon graphs, all of the signs are the same (negative or positive, respectively) and therefore need not be recorded. Thus, the crossing information of the link diagram is encoded in the topology of the all-A or all-B ribbon graphs. Because of this, the all-A and all-B ribbon graphs (and Turaev surfaces) are especially useful in knot theory.

The Seifert Ribbon Graph: If D is an oriented link diagram, oriented link diagram on a surface, or oriented virtual link diagram, then the Seifert ribbon graph, $\mathbb{S}(D)$, is obtained by choosing the signed arrow marked smoothing that is consistent with the orientation of the link, as in Fig. 5.16. The signed arrow marked smoothings in Fig. 5.16 join each incoming arc with the adjacent outgoing arc.

An example of $\mathbb{S}(D)$ is given in Fig. 5.15d, e. (It is a coincidence that Fig. 5.15c, e have the same underlying ribbon graph).

5.3.2.1   Relating the Signed Ribbon Graphs of a Link Diagram

Every link diagram gives rise to a set $\mathscr{G}(D)$ of signed ribbon graphs. It turns out that all of the signed ribbon graphs of a link diagram are partial duals of each other.

Theorem 5.7. *Let D be a classical link diagram, a link diagram on a surface, or a virtual link diagram. Then the signed ribbon graphs of D are partial duals of each other. That is*

$$G, H \in \mathscr{G}(D) \iff G^{\delta(A)} = H, \text{ for some } A \subseteq E(G).$$

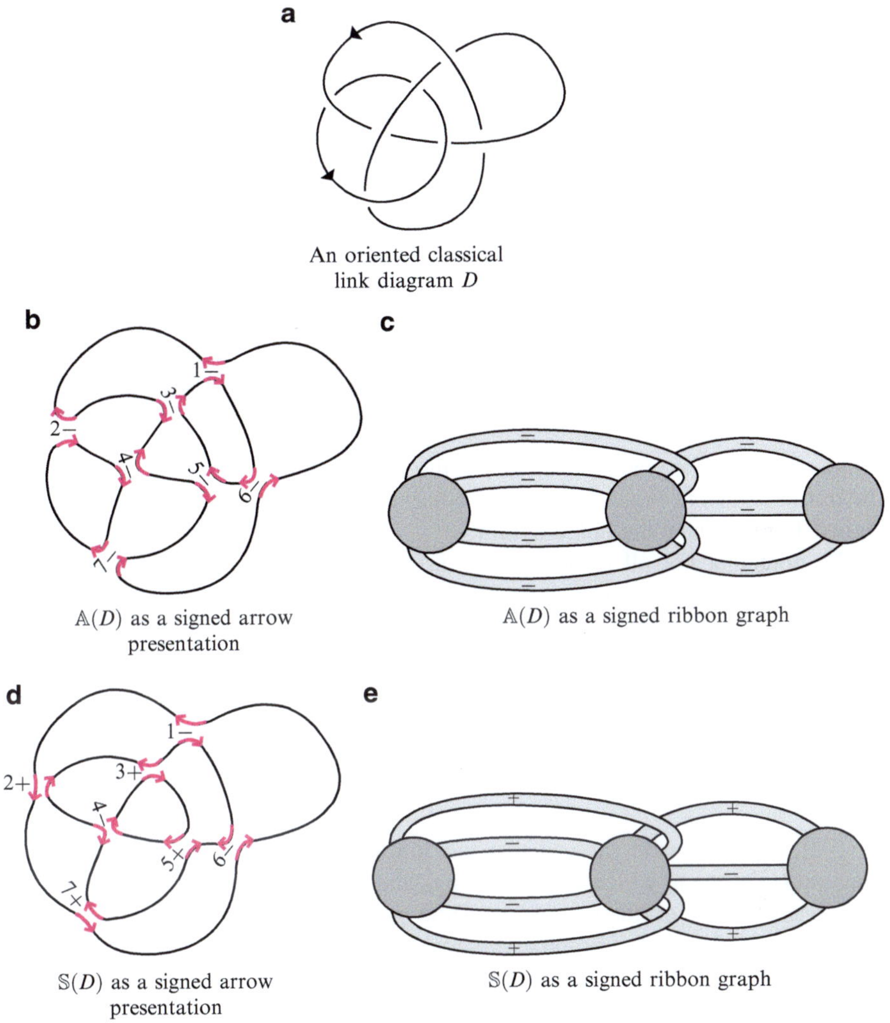

a

An oriented classical
link diagram D

b

$\mathbb{A}(D)$ as a signed arrow
presentation

c

$\mathbb{A}(D)$ as a signed ribbon graph

d

$\mathbb{S}(D)$ as a signed arrow
presentation

e

$\mathbb{S}(D)$ as a signed ribbon graph

Fig. 5.15 Some signed ribbon graphs of the classical link diagram D

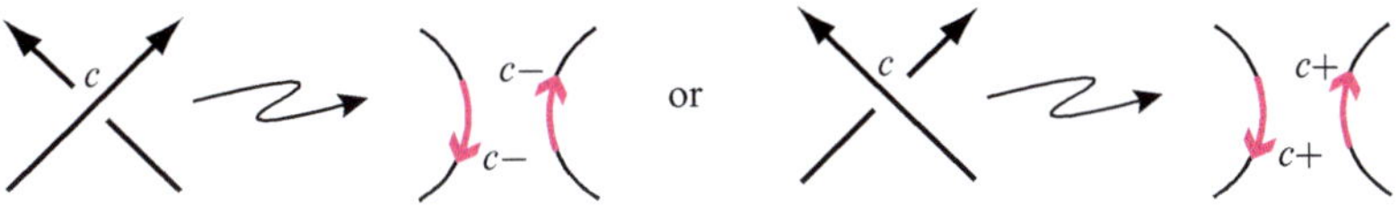

Fig. 5.16 Choosing the signed arrow marked states for a Seifert ribbon graph

Proof. The theorem follows upon observing that changing the choice of an arrow
marked smoothing at a vertex (see Fig. 5.11) corresponds exactly to the change in
an arrow presentation obtained by forming the partial dual at that edge (see Fig. 2.4)

and recalling that forming the dual at an edge of a signed ribbon graph changes the sign of that edge.   □

When a link diagram is checkerboard colourable we can form its signed Tait graph. Since the signed Tait graphs appear in $\mathscr{G}(D)$, it follows that the signed ribbon graphs of a checkerboard colourable link diagram are precisely the partial duals of its signed Tait graphs:

Corollary 5.8. *Let D be a checkerboard colourable classical link diagram, virtual link diagram, or link diagram on a surface. Then G is a signed ribbon graph of D if and only if it is a partial dual of its signed Tait graphs. That is*

$$\mathscr{G}(D) = \{\mathbb{T}(D)^{\delta(A)} \,|\, A \subseteq E(\mathbb{T}(D))\},$$

where $\mathbb{T}(D)$ is either of the signed Tait graphs of D.

As an example, it is readily verified that the signed ribbon graphs of the diagram of the trefoil shown in Fig. 5.13 form the complete set of partial duals of either of the signed Tait graphs of the link diagram.

We note that not all ribbon graphs arise as the ribbon graphs of a classical link diagram (since not every connected embedded graph has a plane partial dual). However, a characterisation of the embedded graphs that represent link diagrams was given in [79].

5.4   The Jones Polynomial and Graph Polynomials

In [92], Thistlethwaite showed that the Jones polynomial of an alternating link diagram (i.e., one in which undercrossings and overcrossings alternate along each component) is an evaluation of the Tutte polynomial of a related plane graph (namely an unsigned Tait graph). This result, known as Thistlethwaite's Theorem, provides an important and seminal link between knot theory and graph theory. There have been several recent results that extend Thistlethwaite's Theorem. These results provide connections between the Jones polynomial and the ribbon graph polynomial $R(G)$. The first results in this direction are due to Chmutov and Pak [17], who showed that the Jones polynomial of a checkerboard colourable virtual link diagram (and of a link diagram on a surface) can be recovered as an evaluation of the signed ribbon graph polynomial of its signed Tait graph. Thistlethwaite's Theorem was extended to non-alternating classical links by Dasbach et al. in [21]. They showed that the Jones polynomial is an evaluation of the (non-signed) ribbon graph polynomial of the all-A signed ribbon graph of a link diagram. These connections between graph and knot polynomials were first unified in [77]. Chmutov and Voltz in [18] went on to extend Thistlethwaite's Theorem and the work of Dasbach et al. by showing that the Jones polynomial of a virtual link is an evaluation of the signed ribbon graph polynomial of its Seifert ribbon graph. Chmutov in [16] then showed

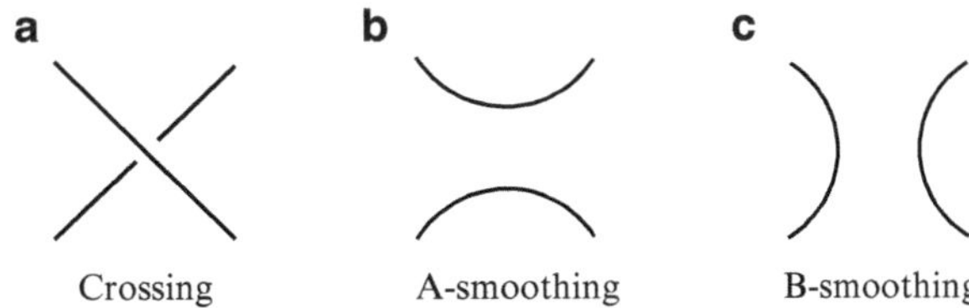

Fig. 5.17 The A-smoothing and B-smoothing of a crossing

that the Jones polynomial of any classical or virtual link can be obtained as an evaluation of the signed ribbon graph polynomial of any of the signed ribbon graphs of a link diagram. These results were extended by Bradford et al. in [8] to include Kauffman and Dye's arrow polynomial (see [25]) of a virtual link. In this section we provide an overview of these connections among the Jones, Tutte, and ribbon graph polynomials.

5.4.1   The Jones Polynomial and the Kauffman Bracket

We approach the Jones polynomial [52] through the Kauffman bracket. The Kauffman bracket is a regular isotopy invariant (so it is not invariant under the Reidemeister type I move). It is defined as a state sum over smoothings of a link diagram. The Kauffman bracket was introduced for classical links by Kauffman in [54] and was quickly extended to links in thickened surfaces (see [20, 46, 47, 69, 86, 94]) and to virtual links (see [55]).

Let D be a classical link diagram, a virtual link diagram, or a link diagram on a surface. A *smoothing* of a (classical) crossing of D is the replacement of that crossing with one of the two schemes shown in Fig. 5.17. Smoothings fall into two types, *A-smoothings* and *B-smoothings*, as defined in Fig. 5.17. As usual, the diagram is unchanged outside of the region shown. These smoothings are the same as those in Fig. 5.11, except without the arrow markings.

A *state* of D is the result of smoothing at each of its crossings. We denote the set of states of D by $\mathscr{S}(D)$. Observe that if D is a classical link diagram or a link diagram on a surface, then a state is a collection of closed curves embedded in the surface; and if D is a virtual link diagram, it is a collection of closed curves immersed in the plane. These closed curves are called *cycles*.

Definition 5.9. Let D be either a classical link diagram, a link diagram on a surface, or a virtual link diagram. Then the *Kauffman bracket*, $\langle D \rangle \in \mathbb{Z}[A, A^{-1}]$, is the state sum

$$\langle D \rangle := \sum_{\sigma \in \mathscr{S}(D)} A^{\alpha(\sigma)} A^{-\beta(\sigma)} d^{c(\sigma)-1},$$

where $\alpha(\sigma)$ is the number of A-smoothings in the state σ, the number of B-smoothings in σ is $\beta(\sigma)$, the number cycles in σ is $c(\sigma)$, and $d = -A^2 - A^{-2}$.

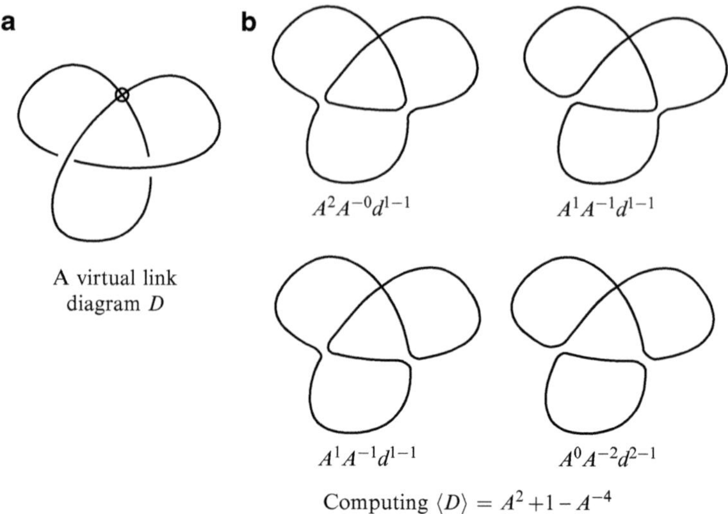

Fig. 5.18 The Kauffman bracket of a virtual link diagram

Fig. 5.19 The oriented signs
of a link diagram

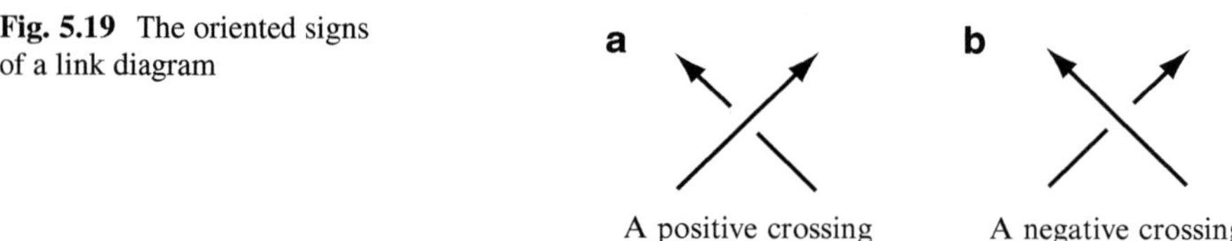

An example of calculating the Kauffman bracket of a virtual link is given in
Fig. 5.18.

The Kauffman bracket can also be defined recursively using the following three
relations:

1. $\left\langle \vcenter{\hbox{$\times$}} \right\rangle = A \left\langle \asymp \right\rangle + A^{-1} \left\langle \,\right)\left(\, \right\rangle$
2. $\langle D \sqcup O \rangle = d \langle D \rangle$
3. $\langle O \rangle = 1$

Here O denotes a diagram with no classical crossings (this need not represent the
unknot in the non-classical case).

As mentioned above, the Kauffman bracket is *not* a knot invariant, but rather
an invariant of regular isotopy. That is, it detects the Reidemeister I move, but it
does not detect the other classical and generalised Reidemeister moves. To obtain
a polynomial that is invariant under the Reidemeister I move, we normalise the
Kauffman bracket using signs of the crossings of an oriented link diagram. The
writhe, $w(D)$, of a link diagram D is its number of positive crossings minus its
number of negative crossings. Here the sign is the *oriented sign* shown in Fig. 5.19.

Fig. 5.20 Link diagrams which are identical except in the region shown

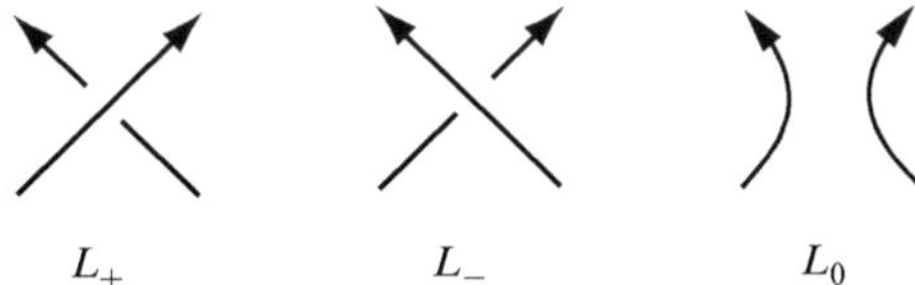

Theorem 5.10. *Let L be a classical link, a link in a thickened surface, or a virtual link, and let D be any diagram of L. Then*

1. *$\langle D \rangle$ is invariant under regular isotopy*
2. *And if L is oriented, then*

$$(-A)^{-3w(D)}\langle D \rangle$$

   is an invariant of L

The virtual knot polynomial that arises in Item 2 of Theorem 5.10 is also known as the *f-polynomial*. For example, the *f*-polynomial of the link shown in Fig. 5.18 equipped with either orientation is $-A^{-4} - A^{-6} + A^{-10}$.

We will now use the Kauffman bracket to define the Jones polynomial.

Definition 5.11. Let L be an oriented classical link, an oriented link in a thickened surface, or an oriented virtual link. Then the *Jones polynomial*, $V(L) \in \mathbb{Z}[t^{-1/2}, t^{1/2}]$, is the Laurent polynomial defined by

$$V(L) = (-A)^{3w(D)}\langle D \rangle \Big|_{A^2 = t^{-1/2}},$$

where D is any diagram of L.

Theorem 5.12. *The Jones polynomial is an invariant of oriented classical links, oriented links in thickened surfaces, and of oriented virtual links. Moreover, it satisfies the following skein relation:*

1. *If L_+, L_-, and L_0 are three links which have diagrams that are identical except in one region where they differ as shown in Fig. 5.20, then*

$$t^{-1}V(L_+) - tV(L_-) = (t^{1/2} - t^{-1/2})V(L_0).$$

2. *$V(\mathcal{O}) = 1$, where $\mathcal{O}$ is the unknot.*

The relations in Theorem 5.12 define the Jones polynomial of classical links, but not of links in thickened surfaces or virtual links (as, in these cases, the first identity in the theorem need not reduce a link to a linear combination of unknots). We discuss this point further in Sect. 5.5.

5.4.2 The Jones Polynomial as a Graph Polynomial

Here we give a unified description of all of the connections between the knot and graph polynomials so far mentioned. We begin by relating the Kauffman bracket and the ribbon graph polynomial of the all-A signed ribbon graph (this result is due to Dasbach et al. [21] for classical links and Chmutov [16] for the other types of links). We then recover all of the other connections from this by using partial duality.

Theorem 5.13 (Dasbach et al. [21]; Chmutov [16]). *Let D be either a classical link diagram, a link diagram on a surface, or a virtual link diagram, and let $\mathbb{A} := \mathbb{A}(D)$ be the all-A ribbon graph of D. Then*

$$\langle D \rangle = d^{k(\mathbb{A})-1}A^{n(\mathbb{A})-r(\mathbb{A})}R(\mathbb{A}; -A^4, A^{-2}d, d^{-1}, 1),$$

where $d = -A^2 - A^{-2}$.

Proof. Let D be either a classical link diagram, a link diagram on a surface, or a virtual link diagram. Obtain a drawing of an arrow presentation for $\mathbb{A}$ on the surface by replacing each crossing c with an arrow marked A-smoothing as in Fig. 5.21a, b. Then we can draw the medial graph $\mathbb{A}_m$ in such a way that it lies on top of D as in Fig. 5.21a, c. Consider Fig. 5.21d, which shows the vertex v_c of $\mathbb{A}_m$ embedded in $\mathbb{A}$. From this figure we see that the arcs of a black smoothing at v_e follow the arcs of the A-smoothing of the crossing c of D, and the arcs of a white smoothing at v_e follow the arcs of the B-smoothing of the crossing c of D (where $\mathbb{A}_m$ has the canonical checkerboard colouring). From this it is easily seen that

$$\langle D \rangle = d^{-1}Q(\mathbb{A}; (\mathbf{A^{-1}}, \mathbf{A}, \mathbf{0}), d) = d^{-1}A^{v(\mathbb{A}_m)}Q(\mathbb{A}; (\mathbf{A^{-2}}, \mathbf{1}, \mathbf{0}), d). \tag{5.1}$$

From Proposition 4.40 it then follows that

$$\langle D \rangle = d^{-1}A^{e(\mathbb{A})}Z(\mathbb{A}; 1, A^{-2}, d, 1), \tag{5.2}$$

which, by Eq. (4.18), gives that

$$\begin{aligned}\langle D \rangle &= d^{k(\mathbb{A})-1}A^{e(\mathbb{A})-2v(\mathbb{A})+2k(\mathbb{A})}R(\mathbb{A}; (d+A^{-2})/A^{-2}, A^{-2}d, 1/d, 1)\\ &= d^{k(\mathbb{A})-1}A^{n(\mathbb{A})-r(\mathbb{A})}R(\mathbb{A}; -A^4, A^{-2}d, d^{-1}, 1),\end{aligned}$$

as required. $\square$

Corollary 5.14. *Let L be an oriented classical link, an oriented link in a thickened surface, or an oriented virtual link, and let $\mathbb{A} = \mathbb{A}(D)$ be the all-A signed ribbon graph of a diagram D of L. Then*

$$V(L;t) = (-1)^{w(D)}t^{(3w(D)-r(\mathbb{A})+n(\mathbb{A}))/4}(-t^{1/2}-t^{-1/2})^{k(\mathbb{A})-1}$$

$$R(\mathbb{A}; -t, -t^{-1}-1, 1/(-t^{1/2}-t^{-1/2}), 1).$$

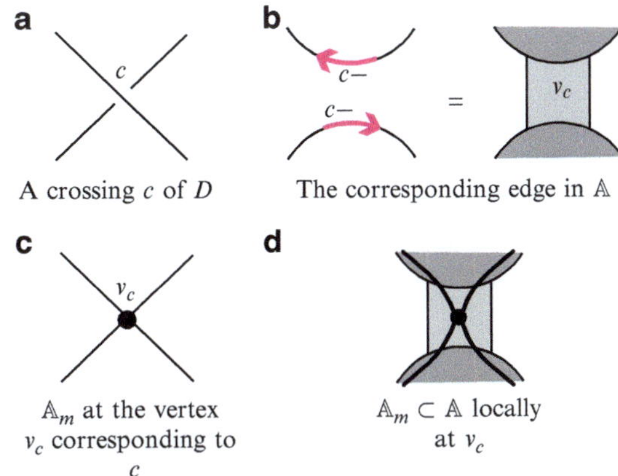

Fig. 5.21 Figures used in the proof of Theorem 5.13

The following result is *Thistlethwaite's Theorem*. It states that the Jones polynomial of an alternating link is an evaluation of the Tutte polynomial of a plane graph.

Corollary 5.15 (Thistlethwaite [92]). *Let L be an alternating oriented classical link, and let $\mathbb{A} = \mathbb{A}(D)$ be the all-A ribbon graph of an alternating diagram D of L. Then*

$$V(L;t) = (-1)^{w(D)} t^{(3w(D)-r(\mathbb{A})+n(\mathbb{A}))/4} (-t^{1/2} - t^{-1/2})^{k(\mathbb{A})-1} T(\mathbb{A}; -t, -t^{-1}).$$

Moreover, $\mathbb{A}$ is a signed Tait graph of D.

Proof. The corollary follows from Corollary 5.14 and Proposition 4.33 upon noting that, since D is alternating and classical, $\mathbb{A}$ is of genus zero. $\qquad\square$

The following result extends Theorem 5.13 to all of the signed ribbon graphs of a link diagram.

Theorem 5.16 (Chmutov [16]). *Let D be either a classical link diagram, a link diagram on a surface, or a virtual link diagram, and let $G \in \mathcal{G}(D)$ be any signed ribbon graph of D. Then*

$$\langle D \rangle = d^{-1} A^{e_-(G)-e_+(G)} Z(G; 1, \mathbf{w}, d, 1),$$

where $d = -A^2 - A^{-2}$, and the weight system $\mathbf{w}$ is given by

$$w_e = \begin{cases} A^{-2} & \text{if e is negative,} \\ A^2 & \text{if e is positive.} \end{cases}$$

Proof. Starting with Eq. (5.2) and using the weight system from the statement of the theorem, we can write

$$\langle D \rangle = d^{-1}A^{e(\mathbb{A})}Z(\mathbb{A};1,A^{-2},d,1) = d^{-1}A^{e(\mathbb{A})}Z(\mathbb{A};1,\mathbf{w},d,1),$$

where the second equality follows since every edge in $\mathbb{A}$ is negative. By Theorem 5.7, we have $G^{\delta(X)} = \mathbb{A}$, where X is the set of positive edges of G. Then, by Theorem 4.44, we have

$$d^{-1}A^{e(\mathbb{A})}Z(\mathbb{A};1,\mathbf{w},d,1) = d^{-1}A^{e(\mathbb{A})}\left(\prod_{e\in X}w_e\right)^{-1}Z(\mathbb{A};1,\mathbf{w},d,1)$$

$$= d^{-1}A^{e(G)}A^{-2|X|}Z(G;1,\mathbf{w},d,1)$$

$$= d^{-1}A^{e(G)}A^{-2e_+(G)}Z(G;1,\mathbf{w},d,1)$$

$$= d^{-1}A^{e_-(G)-e_+(G)}Z(G;1,\mathbf{w},d,1). \qquad \square$$

We note that restricting Theorem 5.16 to classical links gives Kauffman's connection between the dichromatic polynomial and the Kauffman bracket from [53].

Corollary 5.17. *Let D be either a classical link diagram, a link diagram on a surface, or a virtual link diagram, and let $\mathbb{B} = \mathbb{B}(D)$ be the all-B signed ribbon graph of D. Then*

$$\langle D \rangle = d^{k(\mathbb{B})-1}A^{r(\mathbb{B})-n(\mathbb{B})}R(\mathbb{B};-A^{-4},A^2d,d^{-1},1),$$

where $d = -A^2 - A^{-2}$.

Corollary 5.18. *Let L be an oriented classical link, an oriented link in a thickened surface, or an oriented virtual link, and let $\mathbb{B} = \mathbb{B}(D)$ be the all-B signed ribbon graph of a diagram D of L. Then*

$$V(L;t) = (-1)^{w(D)}t^{(3w(D)-n(\mathbb{B})+r(\mathbb{B}))/4}(-t^{1/2}-t^{-1/2})^{k(\mathbb{B})-1}$$

$$\cdot R(\mathbb{B};-t^{-1},-t-1,1/(-t^{1/2}-t^{-1/2}),1).$$

Corollary 5.19 (Thistlethwaite [92]). *Let L be an alternating oriented classical link, and let $\mathbb{B} = \mathbb{B}(D)$ be the all-B ribbon graph of an alternating diagram D of L. Then*

$$V(L;t) = (-1)^{w(D)}t^{(3w(D)-n(\mathbb{B})+r(\mathbb{B}))/4}(-t^{1/2}-t^{-1/2})^{k(\mathbb{B})-1}T(\mathbb{B};-t^{-1},-t^1).$$

Moreover, $\mathbb{B}$ is a signed Tait graph of D.

For comparing Corollary 5.20 with the literature, we reiterate that Chmutov in [16] uses a different sign convention in his ribbon graphs of a link diagram from that used here.

Corollary 5.20 (Chmutov [16]). *Let D be either a classical link diagram, a link diagram on a surface, or a virtual link diagram, and let $G \in \mathcal{G}(D)$ be any signed ribbon graph of D. Then*

$$\langle D \rangle = d^{k(G)-1} A^{r(G)-n(G)} R_s(G; A^{-2}d, A^2d, d^{-1}),$$

where $d = -A^2 - A^{-2}$ and $R_s(G)$ is the signed ribbon graph polynomial from Sect. 4.5.4.

Proof. Starting with Theorem 5.16, we have

$$\langle D \rangle = d^{-1} A^{e_-(G)-e_+(G)} Z(G; 1, \mathbf{w}, d, 1)$$
$$= d^{-1} A^{e_-(G)-e_+(G)} (dA^{-2})^{k(G)} A^{2v(G)} A^{-2e_-(G)} R_s(G; A^{-2}d, A^2d, d^{-1})$$
$$= d^{k(G)-1} A^{r(G)-n(G)} R_s(G; A^{-2}d, A^2d, d^{-1}),$$

where the second equality follows by Eq. (4.27). $\qquad\square$

We note that in the case when D is checkerboard colourable and virtual, and G is its signed Tait graph, Corollary 5.20 is the main result of [17]. Also, when Corollary 5.20 is applied to the Seifert ribbon graph of a virtual link diagram, Corollary 5.20 is the main result of [18].

5.4.3 The Kauffman Bracket and the Transition Polynomial

The signed ribbon graphs of a link diagram are constructed from a link diagram by replacing each crossing with an arrow marked smoothing as in Fig. 5.11. By thinking of a link diagram as a 4-regular embedded graph with decorated vertices, we can regard the signed ribbon graphs of a link diagram as signed smoothing graphs (Definition 3.4) of the 4-regular embedded graph. Recalling that smoothing graphs are special cycle family graphs, it is then natural to ask for the description of link diagrams by embedded graphs that corresponds to the full set of cycle family graphs, rather than just the smoothing graphs. Then once we have this set of graphs, we want an analogue of Theorem 5.16 which expresses the Kauffman bracket as evaluations of the polynomials of the signed ribbon graphs of a link. In this section, we describe such an extension of the signed ribbon graphs of a link and describe the Kauffman bracket in terms of the topological transition polynomial of these ribbon graphs.

Let D be a classical link diagram, a link diagram on a surface, or a virtual link diagram. Assign a unique label to each crossing of D. A *weighted vertex state* of D is the replacement of the crossing with one of the six schemes shown in Fig. 5.22. A *weighted state s* of a link diagram is an assignment of a weighted vertex state to each crossing. Observe that a weighted state is precisely a weighted arrow presentation.

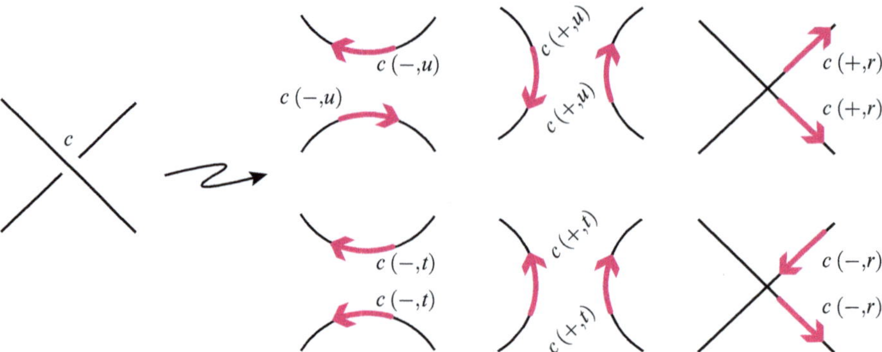

Fig. 5.22 The weighted vertex states of a link diagram

Definition 5.21. Let D be a classical link diagram, a link diagram on a surface, or a virtual link diagram. In addition let s be a weighted state of D. Then the weighted ribbon graph corresponding to the weighted arrow presentation s is a *weighted ribbon graph of the link diagram* D.

An example of the weighted ribbon graph of the link diagram is given in Fig. 5.23.

The following theorem extends Theorem 5.16 to the weighted ribbon graphs of a link diagram.

Theorem 5.22. *Let D be either a classical link diagram, a link diagram on a surface, or a virtual link diagram, and let G be any weighted ribbon graph of D. Then*

$$\langle D \rangle = d^{-1} Q(G; W_{\langle D \rangle}, d),$$

where $d = -A^2 - A^{-2}$, and the weight system $W_{\langle D \rangle}$ is obtained by using the weights on the edges of G to determine the vertex state weights as in the following table.

Edge weight	$(-,u)$	$(-,t)$	$(-,r)$
Vertex state weights	$(A^{-1},A,0)$	$(0,A,A^{-1})$	$(A^{-1},0,A)$
Edge weight	$(+,u)$	$(+,t)$	$(+,r)$
Vertex state weights	$(A,A^{-1},0)$	$(0,A^{-1},A)$	$(A,0,A^{-1})$

Proof. By regarding D as a decorated 4-regular embedded graph, it follows from Theorem 3.8 that the unweighted ribbon graph G is a twisted dual of $\mathbb{A}$. Specifically, if we let $E_{(\pm,x)}$ denote the set of $(\pm,x)$-weighted edges of G, we have $G = \mathbb{A}^{\Gamma}$, where

$$\Gamma = 1(E_{(-,u)})\delta(E_{(+,u)})\tau(E_{(-,t)})\delta\tau(E_{(+,r)})\tau\delta(E_{(+,t)})\tau\delta\tau(E_{(-,r)}).$$

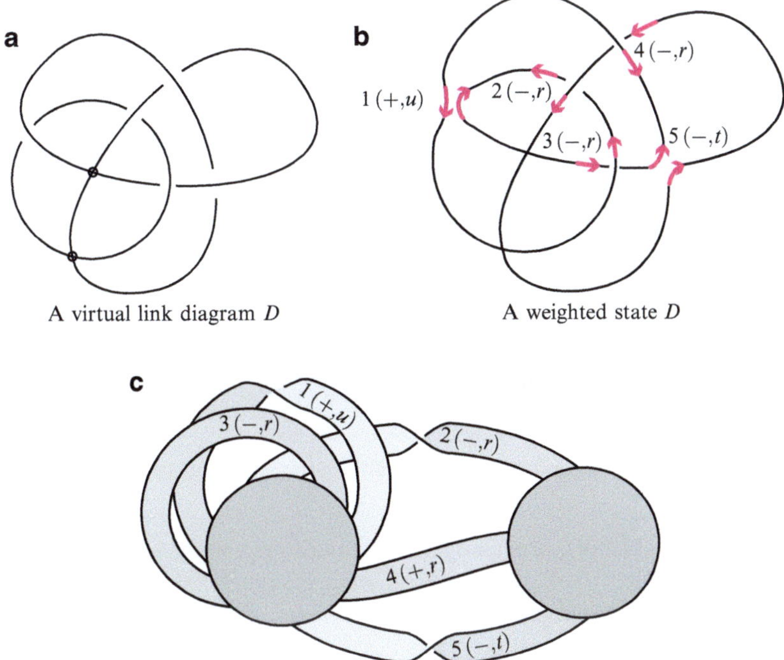

Fig. 5.23 A weighted ribbon graph of a link diagram

Then, by Eq. (5.1) and Theorem 4.12,

$$\langle D \rangle = d^{-1}Q(\mathbb{A};(\mathbf{A}^{-1},\mathbf{A},\mathbf{0}),d) = d^{-1}Q(\mathbb{A}^{\Gamma};(\mathbf{A}^{-1},\mathbf{A},\mathbf{0})^{\Gamma},d),$$

and the result follows upon observing that $(\mathbf{A}^{-1},\mathbf{A},\mathbf{0})^{\Gamma} = W_{\langle D \rangle}$. $\qquad\qquad$ □

5.5   The HOMFLY-PT Polynomial and Graph Polynomials

In Sect. 5.4 it was shown that the Jones polynomial can be obtained as an evaluation of the ribbon graph polynomial. In this section we ask what knot polynomial determines the ribbon graph polynomial. We will see that along the surface $xyz^2 = 1$, the ribbon graph polynomial is determined by the HOMFLY-PT polynomial of links in thickened surfaces. This result from [75] extends a theorem of Jaeger [48], which shows that the Tutte polynomial of a plane graph can be recovered from the HOMFLY-PT polynomial of a classical link. Jaeger's result was extended by Traldi in

[93]. By considering a larger class of links, Traldi was able to relate the HOMFLY-PT polynomial and the dichromatic polynomial. This was extended to the ribbon graph polynomial in [75]. Jaeger and Traldi's work was generalised by Jin and Zhan in [51]. They showed that a large class of links give rise to connections, of the type described by Theorem 5.26, between the dichromatic polynomial of plane graphs and the HOMFLY-PT polynomial of classical links.

In this section we give an overview of Jaeger's relation between the Tutte and HOMFLY-PT polynomials and its extension to the ribbon graph polynomial. We do not include details of Traldi's or Jin and Zhan's extensions here, referring the reader to [51, 75, 93] for these.

5.5.1   The HOMFLY-PT Polynomial

The HOMFLY-PT polynomial is a two-variable generalisation of the Jones polynomial.

Definition 5.23 ([37,87]). Recalling L_+, L_-, and L_0 from Fig. 5.20, the *HOMFLY-PT polynomial*, $P(L;x,y) \in \mathbb{Z}[x,x^{-1},y,y^{-1}]$, of a classical link L in S^3 is defined recursively using the skein relation

$$xP(L_+) - x^{-1}P(L_-) = yP(L_0), \tag{5.3}$$

and its value on the unknot

$$P(\mathcal{O}) = 1. \tag{5.4}$$

Observe that the Jones polynomial of a classical link (which can be obtained from the skein relation in Theorem 5.12) is an evaluation of the HOMFLY-PT polynomial:

$$V(L;t) = P(L;t^{-1}, t^{1/2} - t^{-1/2}).$$

For classical links in S^3, the relations (5.3) and (5.4) define a polynomial since the skein relation (5.3) will reduce any link diagram on S^2 to a $\mathbb{Z}[x^{\pm 1}, y^{\pm 1}]$-linear combination of unknots. However, for link diagrams on an arbitrary oriented surfaces, Eqs. (5.3) and (5.4) are not enough to define the HOMFLY-PT polynomial. For example, consider the link diagram D on a torus T^2 that consists of a simple closed curve that defines a meridian. D is not the diagram of an unknot (as it is the diagram of a link that defines a non-trivial element of $H_1(T^2 \times I)$ and so does not bound a disc). Nor will the skein relation (5.3) reduce it to a linear combination of unknots. Thus Eqs. (5.3) and (5.4) alone are not enough to define a knot invariant. To define the HOMFLY-PT polynomial of a link in a thickened surface we need the concept of a descending link.

There is a natural product of links in $\Sigma \times I$ given by reparameterising the two copies of $\Sigma \times I$ and stacking them:

$$(\Sigma \times I) \times (\Sigma \times I) \to (\Sigma \times [0,1/2]) \times (\Sigma \times [1/2,1]) \to (\Sigma \times I).$$

Denote the projections from $\Sigma \times I$ to Σ and to I by p_Σ and p_I, respectively. The value $p_I(x)$ is called the *height* of x.

Definition 5.24.

1. A knot $K \subset \Sigma \times I$ is *descending* if it is equivalent to a knot $K' \subset \Sigma \times I$, which has the following property: there is a choice of basepoint a on K' such that if we travel along K' in the direction of the orientation from the basepoint, the height of K' decreases until we reach a point a' with $p_\Sigma(a) = p_\Sigma(a')$ from which K' leads back to a by increasing the height and keeping the projection onto Σ constant.
2. A link $L \subset \Sigma \times I$ is said to be *descending* if it is equivalent to a product of descending knots.

Descending links play the role of the unknot in the definition of the HOMFLY-PT polynomial of links in thickened surfaces.

Each equivalence class of a descending knot uniquely determines a conjugacy class of the fundamental group $\pi_1(\Sigma)$. In fact, there is a bijection between the conjugacy classes in $\pi_1(\Sigma)$ and isotopy classes of descending knots, i.e., a conjugacy class determines a descending knot. In [69], Lieberum gives a procedure for choosing a canonical representative of the conjugacy classes of $\pi_1(\Sigma)$. We denote this set of representatives by $\mathscr{B}(\Sigma)$ and associate a commuting indeterminate t_w with each $w \in \mathscr{B}(\Sigma)$. As we will not use it, we will not describe this procedure here. Each descending knot K therefore determines a unique conjugacy class in $\mathscr{B}(\Sigma)$ and therefore an indeterminate t_K. As a descending link is a product of descending knots we can then associate a monomial t_L in commuting indeterminates $\{t_w\}$ to it. For a descending link L,

$$P(L) := t_L \left(\frac{x - x^{-1}}{y} \right)^{k(L)-1}, \tag{5.5}$$

where $k(L)$ is the number of components of the link L.

Definition 5.25.   A *HOMFLY-PT polynomial*, $P(L) \in \mathbb{Z}[x^{\pm 1}, y^{\pm 1}] \otimes \mathbb{Z}[\{t_w\}]$, of a link L in $\Sigma \times I$ is defined by the skein relation

$$xP(L_+) - x^{-1}P(L_-) = yP(L_0), \tag{5.6}$$

and its values on descending links,

$$P(L) = t_L \left(\frac{x - x^{-1}}{y} \right)^{k(L)-1}. \tag{5.7}$$

Here for simplicity and since it is sufficient for our purposes, we will set each $t_w = 1$, obtaining a knot invariant in $\mathbb{Z}[x^{\pm 1}, y^{\pm 1}]$. We denote this invariant by $P(L; x, y)$ and call it *the HOMFLY-PT polynomial* of a link in a thickened surface.

5.5.2   *Graph Polynomials from the HOMFLY-PT Polynomial*

To show that the ribbon graph polynomial arises as an evaluation of the HOMFLY-PT polynomial, we need to associate an oriented link $\mathscr{L}(G)$ to a ribbon graph G. To do this we view ribbon graphs as cellularly embedded graphs. Let G be a cellularly embedded graph in a surface Σ. Form a link diagram on Σ by positioning a special tangle over each edge as indicated in Fig. 5.24 and then connecting the ends of the tangles by following the facial boundaries of G. Let $\mathscr{L}(G) \subset \Sigma \times I$ be a link with this diagram. An example is shown in Fig. 5.25.

The construction of an oriented link $\mathscr{L}(G)$ from a plane graph G as in Fig. 5.24 is due to Jaeger (see [48]). Jaeger used this construction to connect the Tutte polynomial and the HOMFLY-PT polynomial (see Corollaries 5.29 and 5.31 below). The choice of tangle used in the construction of $\mathscr{L}(G)$ is by no means unique, and other tangles have been used by Traldi in [93] (see also [75, 76] for the non-plane case) and by Jin and Zhan in [51].

The following theorem relates the HOMFLY-PT polynomial and $Z(G)$.

Theorem 5.26 ([75]). *Let Σ be an oriented surface and $G \subset \Sigma$ be an embedded graph. Then*

$$P(\mathscr{L}(G);x,y) = \left(\frac{y}{x-x^{-1}}\right)\left(\frac{y}{x}\right)^{e(G)} Z\left(G;1,\frac{1}{xy},\frac{x-x^{-1}}{y}\right).$$

Proof. Rather than working with the links $\mathscr{L}(G)$, we will work with their diagrams as constructed in Fig. 5.24. In this proof we will abuse notation and denote these diagrams also by $\mathscr{L}(G)$. By applying the skein relation of the HOMFLY-PT polynomial [Eq. (5.6)] to the tangle diagram in Fig. 5.24 we obtain

Fig. 5.24   Forming the link $\mathscr{L}(G)$

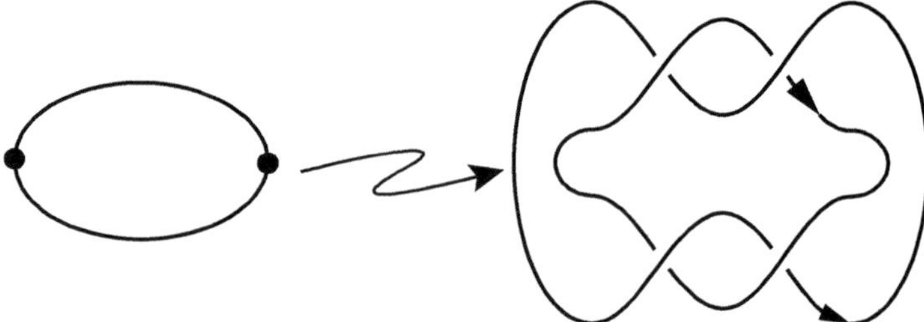

Fig. 5.25   An embedded graph G and a diagram of $\mathscr{L}(G)$

$$P\left(\vcenter{\hbox{(crossing diagram)}} \right) = \frac{1}{x^2} P\left(\vcenter{\hbox{(smoothed diagram)}} \right) + \frac{y}{x} P\left(\vcenter{\hbox{(split diagram)}} \right). \tag{5.8}$$

We can apply Eq. (5.8) to $\mathcal{L}(G)$ to obtain a $\mathbb{Z}[x^{\pm 1}, y^{\pm 1}]$-linear combination of link diagrams that have no crossings. (Note that these are not necessarily diagrams of the unlink.) We will call the link diagrams in this linear combination the *resolutions* of $\mathcal{L}(G)$.

There is a natural bijection φ between the set of spanning subgraphs of G and the set of resolutions of $\mathcal{L}(G)$ which is defined by setting $\varphi(A)$ to be the resolution of $\mathcal{L}(G)$ that bounds the faces of $(V(G), A) \subset \Sigma$. (This means that when $e \in A$, the corresponding tangle is resolved as $\vcenter{\hbox{)(}}$, and when $e \notin A$, the corresponding tangle is resolved as $\vcenter{\hbox{) (}}$.) Under this bijection, $f(A)$ is the number of components of the resolution.

As diagrams with no crossings represent descending links, we have

$$P(\mathcal{L}(G); x, y) = \sum_{A \subseteq E(G)} \left(\frac{1}{x^2} \right)^{e(A)} \left(\frac{y}{x} \right)^{e(G) - e(A)} \left(\frac{x - x^{-1}}{y} \right)^{f(A) - 1}$$

$$= \left(\frac{y}{x - x^{-1}} \right) \left(\frac{y}{x} \right)^{e(G)} \sum_{A \subseteq E(G)} \left(\frac{1}{xy} \right)^{e(A)} \left(\frac{x - x^{-1}}{y} \right)^{f(A)}$$

$$= \left(\frac{y}{x - x^{-1}} \right) \left(\frac{y}{x} \right)^{e(G)} Z\left(G; 1, \frac{1}{xy}, \frac{x - x^{-1}}{y} \right). \qquad \square$$

Theorem 5.26 may be recast in terms of the topological transition polynomial.

Corollary 5.27. *Let Σ be an oriented surface and $G \subset \Sigma$ be an embedded graph. Then*

$$P(\mathcal{L}(G); x, y) = \left(\frac{y}{x - x^{-1}} \right) \left(\frac{y}{x} \right)^{e(G)} Q\left(G; (1/xy, 1, 0), \frac{x - x^{-1}}{y} \right).$$

Proof. We have

$$P(\mathcal{L}(G); x, y) = \left(\frac{y}{x - x^{-1}} \right) \left(\frac{y}{x} \right)^{e(G)} Z\left(G; 1, \frac{1}{xy}, \frac{x - x^{-1}}{y} \right)$$

$$= \left(\frac{y}{x - x^{-1}} \right) \left(\frac{y}{x} \right)^{e(G)} Q\left(G; (1/xy, 1, 0), \frac{x - x^{-1}}{y} \right),$$

where the first identity is from Theorem 5.26 and the second is from Proposition 4.40. $\qquad \square$

In fact, a proof Theorem 5.26 that uses the transition polynomial can be obtained by choosing an appropriate weight system.

For reference later, we use Eq. (4.18) to write Theorem 5.26 in terms of the ribbon graph polynomial.

Corollary 5.28 ([75]). *Let Σ be an oriented surface and $G \subset \Sigma$ be an embedded graph. Then*

$$P(\mathscr{L}(G);x,y) = \left(\frac{1}{xy}\right)^{v(G)-1} \left(\frac{y}{x}\right)^{e(G)} \left(x^2 - 1\right)^{k(G)-1}$$
$$R\left(G;x^2, \frac{x-x^{-1}}{xy^2}, \frac{y}{x-x^{-1}}\right). \tag{5.9}$$

When Σ is S^2, by Proposition 4.33, Theorem 5.28 recovers Jaeger's relation between the HOMFLY-PT polynomial of a classical link and the Tutte polynomial of a plane graph.

Corollary 5.29 (Jaeger [48]). *Let G be a plane graph. Then*

$$P(\mathscr{L}(G);x,y) = \left(\frac{1}{xy}\right)^{v(G)-1} \left(\frac{y}{x}\right)^{e(G)} \left(x^2 - 1\right)^{k(G)-1}$$
$$T\left(G;x^2, \frac{x-x^{-1}}{xy^2} + 1\right). \tag{5.10}$$

Observe that the left-hand and right-hand sides of Eq. (5.10) each contain a 2-variable polynomial. It is natural to ask if these polynomials determine one another. Also, Eq. (5.9) relates the 2-variable HOMFLY-PT polynomial to a 2-variable specialisation of the ribbon graph polynomial. Thus, we are similarly led to ask to what extent the HOMFLY-PT polynomial determines the ribbon graph polynomial. In the following theorem, we will see that the HOMFLY-PT polynomial determines the ribbon graph polynomial $R(G;x+1,y,1/\sqrt{xy},1)$. An important observation is that this specialisation of the ribbon graph polynomial includes that used in obtaining the Kauffman bracket and Jones polynomial (Theorem 5.13 and Corollary 5.14), and it is the specialisation that appears in its duality relation (Theorem 4.45).

Theorem 5.30 ([75]). *If G is an embedded graph, and $\alpha \neq 0, -1$, $\beta \neq 0$, then*

$$R\left(G;\alpha+1,\beta,1/\sqrt{\alpha\beta}\right) = \left(\sqrt{\alpha}/\sqrt{\beta}\right)^{v(G)-e(G)-1} (\alpha+1)^{e(G)}\alpha^{1-k(G)}$$
$$P\left(\mathscr{L}(G);\sqrt{\alpha-1},\sqrt{\frac{\alpha}{(\alpha+1)\beta}}\right). \tag{5.11}$$

Proof. To prove the theorem we need to find non-zero values of x and y such that $\alpha + 1 = x^2$, $\beta = (x-x^{-1})/xy^2$, and $\gamma = y/(x-x^{-1})$. We see that this is possible for the choices $x = \sqrt{\alpha + 1} \neq 1$ and $y = \sqrt{\alpha/(\alpha + 1)\beta}$. The proof of Theorem 5.30 then follows. Notice that $\alpha\beta\gamma^2 = 1$. $\qquad\square$

Corollary 5.31 (Jaeger [48]). *If G is a plane graph, and $x \neq 0, 1$, $y \neq 1$, then*

$$T(G;x,y) = \left(\sqrt{x-1}/\sqrt{y-1}\right)^{v(G)-e(G)-1} x^{e(G)}(x-1)^{1-k(G)}$$

$$P\left(\mathscr{L}(G); \sqrt{x-2}, \sqrt{\frac{x-1}{x(y-1)}}\right). \tag{5.12}$$

It was noted above that the $R(G; x+1, y, 1/\sqrt{xy}, 1)$ is the specialisation that appears in its duality relation (Theorem 4.48). In [75] it was shown that the duality relation for the ribbon graph polynomial is a consequence of the fact that $R(G; x+1, y, 1/\sqrt{xy}, 1)$ is an evaluation of the HOMFLY-PT polynomial. (In fact, the duality relation for the ribbon graph polynomial was discovered independently in [33, 75]. It was discovered in [75] because of its connection with the HOMFLY-PT polynomial.) Moreover, the partial duality relation of Theorem 4.48 can also be recovered from the connections between $R(G)$ and the HOMFLY-PT polynomial (see [76]) providing a good example of the interplay between knot theory and graph theory.

References

1. Abe T (2009) The Turaev genus of an adequate knot. Topology Appl 156:2704–2712
2. Aigner M (1997) The Penrose polynomial of a plane graph. Math Ann 307:173–189
3. Askanazi R, Chmutov S, Estill C, Michel J, Stollenwerk P (2013) Polynomial invariants of graphs on surfaces. Quantum Topol 3:77–90
4. Bollobás B (1998) Modern graph theory. In: Graduate texts in mathematics, vol 184. Springer, New York
5. Bollobás B, Riordan O (2001) A polynomial for graphs on orientable surfaces. Proc London Math Soc 83:513–531
6. Bollobás B, Riordan O (2002) A polynomial of graphs on surfaces. Math Ann 323:81–96
7. Bondy J, Murty U (2008) Graph theory. In: Graduate texts in mathematics, vol 244. Springer, New York
8. Bradford R, Butler C, Chmutov S (2012) Arrow ribbon graphs. J Knot Theory Ramifications 21:1240002
9. Burde G, Zieschang H (2003) Knots, 2nd edn. de Gruyter studies in mathematics, vol 5. Walter de Gruyter, Berlin
10. Brylawski T (1982) The Tutte polynomial. Part 1: General theory. In: Barlotti A (ed) Matroid theory and its applications, proceedings of the third international mathematical summer center (C.I.M.E. 1980), pp 125–275
11. Brylawski T, Oxley J (1992) The Tutte polynomial and its applications. In: White N (ed) Matroid applications, encyclopedia of mathematics and its applications. Cambridge University Press, Cambridge
12. Carter J, Kamada S, Saito M (2002) Stable equivalences of knots on surfaces and virtual knot cobordisms. J Knot Theory Ramifications 11:311–322
13. Champanerkar A, Kofman I (2009) Spanning trees and Khovanov homology. Proc Amer Math Soc 137:2157–2167
14. Champanerkar A, Kofman I, Stoltzfus N (2007) Graphs on surfaces and Khovanov homology. Algebr Geom Topol 7:1531–1540
15. Champanerkar A, Kofman I, Stoltzfus N (2011) Quasi-tree expansion for the Bollobás-Riordan-Tutte polynomial. Bull Lond Math Soc 43:972–984
16. Chmutov S (2009) Generalized duality for graphs on surfaces and the signed Bollobás-Riordan polynomial. J Combin Theory Ser B 99:617–638
17. Chmutov S, Pak I (2007) The Kauffman bracket of virtual links and the Bollobás-Riordan polynomial. Mosc Math J 7:409–418
18. Chmutov S, Voltz J (2008) Thistlethwaite's theorem for virtual links. J Knot Theory Ramifications 17:1189–1198

J.A. Ellis-Monaghan and I. Moffatt, *Graphs on Surfaces: Dualities, Polynomials, and Knots*, SpringerBriefs in Mathematics, DOI 10.1007/978-1-4614-6971-1,
© Joanna A. Ellis-Monaghan, Iain Moffatt 2013

19. Crowell R, Fox R (1977) Introduction to knot theory. In: Graduate texts in mathematics, vol 57. Springer, New York
20. Cotta-Ramusino P, Rinaldi M (1991) On the algebraic structure of link-diagrams on a 2-dimensional surface. Comm Math Phys 138:137–173
21. Dasbach O, Futer D, Kalfagianni E, Lin X-S, Stoltzfus N (2008) The Jones polynomial and graphs on surfaces. J Combin Theory Ser B 98:384–399
22. Dasbach O, Futer D, Kalfagianni E, Lin X-S, Stoltzfus N (2010) Alternating sum formulae for the determinant and other link invariants. J Knot Theory Ramifications 19:765–782
23. Dasbach O, Lowrance A (2011) Turaev genus, knot signature, and the knot homology concordance invariants. Proc Amer Math Soc 139:2631–2645
24. Diestel R (2010) Graph theory. In: Graduate texts in mathematics, vol 173, 4th edn. Springer, Heidelberg
25. Dye H, Kauffman L (2009) Virtual crossing number and the arrow polynomial. J Knot Theory Ramifications 18:1335–1357
26. Ellis-Monaghan J, Merino C (2011) Graph polynomials and their applications I: the Tutte polynomial. In: Dehmer M (ed) Structural analysis of complex networks. Birkhauser, New York, pp 219–255
27. Ellis-Monaghan J, Merino C (2011) Graph polynomials and their applications II: interrelations and interpretations. In: Dehmer M (ed) Structural analysis of complex networks. Birkhauser, New York, pp 257–292
28. Ellis-Monaghan J, Moffatt I (2013) A Penrose polynomial for embedded graphs. Eur J Combin 34:424–445
29. Ellis-Monaghan J, Moffatt I. Evaluations of topological Tutte polynomials. arXiv:1108.3321 (preprint)
30. Ellis-Monaghan J, Moffatt I (2012) Twisted duality and polynomials of embedded graphs. Trans Amer Math Soc 364:1529–1569
31. Ellis-Monaghan J, Sarmiento I (2001) Medial graphs and the Penrose polynomial. Congr Numer, 150:211–222
32. Ellis-Monaghan J, Sarmiento I (2002) Generalized transition polynomials. Congr Numer 155:57–69
33. Ellis-Monaghan J, Sarmiento I. A duality relation for the topological Tutte polynomial, talk at the AMS eastern section meeting special session on graph and matroid invariants, bard college. http://academics.smcvt.edu/jellis-monaghan/#Research. Accessed 10 Sept 2005
34. Ellis-Monaghan J, Sarmiento I (2011) A recipe theorem for the topological Tutte polynomial of Bollobás and Riordan. Eur J Combin 32:782–794
35. Fox R, Artin E (1948) Some wild cells and spheres in three-dimensional space. Ann Math 49:979–990
36. Fox R (1949) A remarkable simple closed curve. Ann Math 50:264–265
37. Freyd P, Hoste J, Lickorish W, Millett K, Ocneanu A, Yetter D (1985) A new polynomial invariant of knots and links. Bull Amer Math Soc (N.S.) 12:239–246
38. Futer D, Kalfagianni E, Purcell J (2008) Dehn filling, volume, and the Jones polynomial. J Differ Geom 78:429–464
39. Futer D, Kalfagianni E, Purcell J (2009) Symmetric links and Conway sums: volume and Jones polynomial. Math Res Lett 16:233–253
40. Goussarov M, Polyak M, Viro O (2000) Finite-type invariants of classical and virtual knots. Topology 39:1045–1068
41. Gross J, Tucker T (1987) Topological graph theory, Wiley-Interscience, New York
42. Gurau R (2010) Topological graph polynomials in colored group field theory. Ann Henri Poincaré 11:565–584
43. Huggett S, Moffatt I (2011) Expansions for the Bollobás-Riordan and Tutte polynomials of separable ribbon graphs. Ann Comb 15:675–706
44. Huggett S, Moffatt I. Bipartite partial duals and circuits in medial graphs, Combinatorica (in press) arXiv:1106.4189

45. Huggett S, Moffatt I, Virdee N (2012) On the Seifert graphs of link diagrams and its parallels. Math Proc Cambridge Philos Soc 153:123–145
46. Hoste J, Przytycki J (1989) An invariant of dichromatic links. Proc Amer Math Soc 105:1003–1007
47. Inoue K, Kaneto T (1994) A Jones type invariant of links in the product space of a surface and the real line. J Knot Theory Ramifications 3:153–161
48. Jaeger F (1988) Tutte polynomials and link polynomials. Proc Amer Math Soc 103:647–654
49. Jaeger F (1988) On Tutte polynomials and cycles of plane graphs. J Combin Theory Ser B 44:127–146
50. Jaeger F (1990) On transition polynomials of 4-regular graphs. In: Hahn G. et al (eds) Cycles and rays (Montreal, PQ, 1987). NATO advanced science institutes series c: mathematical and physical sciences, vol 301. Kluwer Academic, Dordrecht, pp 123–150
51. Jin X, Zhang F (2012) The Homfly and dichromatic polynomials. Proc Amer Math Soc 140:1459–1472
52. Jones V (1985) A polynomial invariant for knots via von Neumann algebras. Bull Amer Math Soc 12:103–111
53. Kauffman L (1987) State models and the Jones polynomial. Topology 26:395–407
54. Kauffman L (1990) An invariant of regular isotopy. Trans Amer Math Soc 312:417–471
55. Kauffman L (1999) Virtual knot theory. Eur J Combin 20:663–690
56. Kauffman L (2000) A survey of virtual knot theory. In: Knots in Hellas '98 (Delphi), Ser. Knots Everything, vol 24. World Science, River Edge, pp 143–202
57. Kauffman L (2012) Introduction to virtual knot theory. J. Knot Theory Ramifications 21:1240007
58. Kauffman L, Manturov V (2006) Virtual knots and links, Tr. Mat. Inst. Steklova Geom. Topol., Diskret. Geom. i Teor. Mnozh., translation in Proc. Steklov Inst. Math. 25:114–133; 252:104–121.
59. Kamada N (2002) On the Jones polynomials of checkerboard colorable virtual links. Osaka J Math 39:325–333
60. Kamada N (2004) Span of the Jones polynomial of an alternating virtual link. Algebr Geom Topol 4:1083–1101
61. Korn M, Pak I. Combinatorial evaluations of the Tutte polynomial (preprint)
62. Kotzig A (1968) Eulerian lines in finite 4-valent graphs and their transformations. Theory of Graphs (Proc. Colloq., Tihany, 1966) Academic, New York, pp 219–230
63. Krajewski T, Rivasseau V, Vignes-Tourneret F (2011) Topological graph polynomials and quantum field theory. Part II: Mehler kernel theories. Ann Henri Poincaré 12:1–63
64. Lando S, Zvonkin A (2004) Graphs on surfaces and their applications, encyclopaedia of mathematical sciences, low-dimensional topology, II. vol 141. Springer, Berlin
65. Las Vergnas M (1981) Eulerian circuits of 4-valent graphs imbedded in surfaces. In: Algebraic methods in graph theory, vols I, II (Szeged, 1978), vol 25. Colloquia Mathematica Societatis Jnos Bolyai, North-Holland, Amsterdam, New York, pp 451–477
66. Las Vergnas M (1978) Sur les activités des orientations d'une géométrie combinatoire. In: colloque mathématiques discrètes: codes et hypergraphes (Brussels, 1978). cahiers centre Études rech Opér 20:293–300
67. Las Vergnas M (1980) On the Tutte polynomial of a morphism of matroids. Ann Discrete Math 8:7–20
68. Lickorish W 1997 An introduction to knot theory. In: Graduate texts in mathematics, vol 175. Springer, New York
69. Lieberum J (2002) Skein modules of links in cylinders over surfaces. Int J Math Math Sci 32:515–554
70. Loebl M, Moffatt I (2008) The chromatic polynomial of fatgraphs and its categorification. Adv Math 217:1558–1587
71. Lowrance A (2008) On knot Floer width and Turaev genus. Algebr Geom Topol 8:1141–1162
72. Manturov V (2004) Knot theory. Chapman & Hall/CRC, Boca Raton

73. Massey W (1977) Algebraic topology: an introduction. In: Graduate texts in mathematics, vol 56. Springer, New York
74. Milnor J (1964) Most knots are wild. Fund Math 54:335–338
75. Moffatt I (2008) Knot invariants and the Bollobás-Riordan polynomial of embedded graphs. Eur J Combin 29: 95–107
76. Moffatt I (2010) Partial duality and Bollobás and Riordan's ribbon graph polynomial. Discrete Math 310:174–183
77. Moffatt I (2011) Unsigned state models for the Jones polynomial. Ann Comb 15: 127–146
78. Moffatt I (2011) A characterization of partially dual graphs. J Graph Theory 67:198–217
79. Moffatt I (2012) Partial duals of plane graphs, separability and the graphs of knots. Algebr Geom Topol 12:1099–1136
80. Moffatt I (2013) Separability and the genus of a partial dual. Eur J Combin 34:355–378
81. Mohar B, Thomassen C (2001) Graphs on surfaces (Johns Hopkins studies in the mathematical sciences). John Hopkins University Press, Baltimore
82. Nelson S (2001) Unknotting virtual knots with Gauss diagram forbidden moves. J Knot Theory Ramifications 10:931–935
83. Oxley J, Welsh D (1979) The Tutte polynomial and percolation. In: Bondy JA, Murty USR (eds) Graph theory and related topics. Academic, London
84. Penrose R (1971) Applications of negative dimensional tensors, combinatorial mathematics and its applications. In: Proceedings of the conference held in Oxford in 1969. Academic, London, pp 221–244
85. Polyak M (2010) Minimal generating sets of Reidemeister moves. Quantum Topol 1:399–411
86. Przytycki J (1999) Fundamentals of Kauffman bracket skein modules. Kobe J Math 16:45–66
87. Przytycki J, Traczyk P (1988) Invariants of links of Conway type. Kobe J Math 4:115–139
88. Rolfsen D (1976) Knots and links. Mathematics lecture series, No. 7. Publish or Perish, Berkeley, California
89. Tanasa A (2011) Generalization of the Bollobás-Riordan polynomial for tensor graphs. J Math Phys 52:073514
90. Tanasa A (2012) Some combinatorial aspects of quantum field theory. Seminaire Lotharingien de Combinatoire, B65g
91. Tanasa A (2012) Multi-orientable group field theory. J Phys A: Math Theor 45:165401
92. Thistlethwaite M (1987) A spanning tree expansion of the Jones polynomial. Topology 26:297–309
93. Traldi L (1989) A dichromatic polynomial for weighted graphs and link polynomials. Proc Amer Math Soc 106:279–286
94. Turaev V (1990) The Conway and Kauffman modules of the solid torus. J Soviet Math 52:2799–2805
95. Turaev V (1987) A simple proof of the Murasugi and Kauffman theorems on alternating links. Enseign Math 33:203–225
96. Vignes-Tourneret F (2009) The multivariate signed Bollobás-Riordan polynomial. Discrete Math 309:5968–5981
97. Vignes-Tourneret F (2011) Non-orientable quasi-trees for the Bollobás-Riordan polynomial. Eur J Combin 32: 510–532
98. West D (1996) Introduction to graph theory (2nd edn). Prentice Hall, Upper Saddle River
99. Widmer T (2009) Quasi-alternating Montesinos links. J Knot Theory Ramifications 18: 1459–1469
100. Wilson S (1979) Operators over regular maps. Pacific J Math 81:559–568

Index

J.A. Ellis-Monaghan and I. Moffatt, *Graphs on Surfaces: Dualities, Polynomials, and Knots*, SpringerBriefs in Mathematics, DOI 10.1007/978-1-4614-6971-1,

MIX
Papier aus verantwortungsvollen Quellen
Paper from responsible sources
FSC® C105338

If you have any concerns about our products,
you can contact us on
ProductSafety@springernature.com

In case Publisher is established outside the EU,
the EU authorized representative is:
**Springer Nature Customer Service Center GmbH
Europaplatz 3, 69115 Heidelberg, Germany**

Printed by Libri Plureos GmbH
in Hamburg, Germany